Lecture Notes
in Control and Information Sciences 239

Editor: M. Thoma

Lecture Notes
in Control and Information Sciences 239

Editor: M. Thoma

Springer
London
Berlin
Heidelberg
New York
Barcelona
Budapest
Hong Kong
Milan
Paris
Santa Clara
Singapore
Tokyo

Qing-Guo Wang, Tong Heng Lee and
Kok Kiong Tan

Finite Spectrum Assignment for Time-Delay Systems

Springer

Authors

Qing-Guo Wang
Tong Heng Lee
Kok Kiong Tan
Department of Electrical Engineering, The National University of Singapore,
10 Kent Ridge Crescent, S (119260) Singapore

ISBN 1-85233-065-1 Springer-Verlag London Berlin Heidelberg

British Library Cataloguing in Publication Data
Wang, Qing-Guo
  Finite spectrum assignment for time-delay systems. -
  (Lecture notes in control and information sciences ; 239)
  1. Control theory   2. Delay lines
  I. Title   II. Lee, Tong H.   III. Tan, Kok Kiong
  629.8'312
ISBN 1852330651

Library of Congress Cataloging-in-Publication Data
Wang, Qing-Guo, 1957-
    Finite spectrum assignment for time-delay systems. /   Qing-Guo Wang,
  Tong Heng Lee, and Kok Kiong Tan.
      p.   cm.   --   (Lecture notes in control and information sciences
  ; 239)
    Includes bibliographical references.
    ISBN 1-85233-065-1   (pbk. : alk. paper)
    1. Feedback control systems.   2. Process Control.   3. Time delay
  systems.   I. Lee, Tong Heng, 1958-   .   II. Tan, Kok Kiong, 1967-
  .   III. Title.   IV. Series.
  TJ216.W36   1998   98-34633
  629.8'3- -dc21   CIP r98

Typesetting: Camera ready by authors
Printed and bound at the Athenæum Press Ltd., Gateshead, Tyne & Wear
69/3830-543210 Printed on acid-free paper

Preface

The presence of considerable time delays in many industrial processes is well recognized and achievable performances of conventional unity feedback control systems are degraded if a process has a relatively large time delay compared to its time constants. In this case, dead time compensation is necessary in order to enhance the performances. The most popular scheme for such compensation is the *Smith Predictor*, but it is unsuitable for unstable or lightly damped processes because the compensated closed-loop system always contains the process poles themselves. An alternative scheme for delay elimination from the closed-loop is the *finite spectrum assignment* (FSA) strategy and it can arbitrarily assign the closed-loop spectrum. One may note that the Smith-Predictor Control can be found in delay systems control books and many process control books, but the FSA control is rarely included in these books. It is therefore timely and desirable to fill this gap by writing a book which gives a comprehensive treatment of the FSA approach. This is useful and worthwhile since the FSA provides not only an alternative way but also certain advantages over the Smith-Predictor.

The book presents the state-of-the-art of the finite spectrum assignment for time-delay systems in frequency domain. It mainly contains those works carried out recently by the authors in this field. Most of them have been published and others are awaiting publication. They are assembled together and reorganized in such a way that the presentation is logical, smooth and systematic. The book covers the whole range of the topic from algorithms development to practical issues relating to the application of FSA such as auto-tuning and performance robustness. Both simulation and real-time implementation results are included for illustration of the FSA methodology. The pre-requisites are minimal and only knowledge of a first undergraduate control course is assumed. The book is thus readily accessible and practically useful for senior undergraduate and graduate students and engineers.

In what follows, the contents of the book will be briefly reviewed. Chapter 1 provides a short introduction to delay systems, the control strategies for these systems and practical issues expected in implementation. The FSA is one such control strategy for these systems and it first originated from Manitius and Olbrot (1979) in the time domain. The frequency domain version was proposed by Ichikawa (1985), but it can only be applied to single variable systems with distinct poles. A subsequent version of the algorithm extends the application to systems with multiple poles. Multivariable FSA is also desirable since many industrial systems are of multivariable nature. In addition, modification of the algorithm is also necessary to achieve asymptotic regulation and tracking as an ordinary FSA system usually has non-zero steady-state error in response to a step set-point change or load disturbance. Chapter 2 of the book is mainly a collection of the basic and extended versions of the FSA algorithms.

Automatic tuning and adaptation of PID controllers have been successfully applied to industrial processes. This adaptive technique is also needed for model-based advanced controllers such as the FSA if the same success is expected for their applications in industry, as time-varying time delays and/or dynamics often occur because of varying flows and/or operating conditions and require frequent re-tuning of the controllers. Chapters 3 and 4 deal with these practical issues expected in the implementation of FSA systems. Chapter 3 is concerned with non-parametric estimation of the process and fitting of transfer function models to the frequency response. Two useful and practical methods are described. The first method applies the basic idea due to Astrom (1982). The basic and now renowned relay feedback for PID auto-tuning is extended to identify one or more general points on the process Nyquist curve in the more interesting range of frequency, in similar efficiency as the basic relay method estimates the process critical point. The second method examined in the chapter is an online frequency response estimation method based on the use of Discrete Fourier Transform applied to transient signals in relay or step tests. Chapter 4 is focused on controller design, auto-tuning and adaptation based on the transfer functions fitted to the frequency response estimation by the methods described in Chapter 3. Applications of the techniques in process control are demonstrated by simulation and real-time implementations in these chapters.

In practical situations, models of industrial processes are never perfect and robustness of FSA systems in the presence of inevitable uncertainty should be addressed. The practical stability of finite spectrum-assigned delay systems is first considered in Chapter 5. It is shown that such systems suffer from a practical stability problem, *i.e.* an infinitesimal perturbation in the process may destabilize a nominally stable system. The conditions for practical stability are related to structural aspects of the process and model. Similarly, the conditions for robust stability and performance robustness for perturbations of arbitrary magnitude are derived. These provide design guidelines in the development of the control system.

This book would not be possible without the help of certain people to whom the authors would like to express their appreciation. In particular, we would like to thank Mr. Bi Qiang and Mr. Zou Biao for their works related to Chapters 3 and 4. The editing process would not have been as smooth without the generous assistance of Mr. Zhang Li, Mr. C. H. Gan and Mr. Zhang Yong.

Contents

Time-Delay Systems

1.1 Introduction

A *time-delay* under the context of process control systems may be defined as *the time interval from the application of a control signal to any observable change in the process variable*. Time delays have always been among the most difficult problems encountered in process control. It could occur for various reasons and in different magnitudes. It may be attributed to a delay in measurement of the process variable. One such example is the steel rolling process where the measurement point is located at some distance downstream of the steel press so that there is always a time delay in the feedback of plate thickness to the press controller. The time delay could be due to a delay in control, typical of processes involving transportation of materials in pipelines and conveyor systems such as thermal, composition and distillation processes. For these processes, the actuating devices are physically limited, so that instantaneous impulse-like action is not realizable. In modern digital control systems, time-delay can also arise in the form of sampling intervals, or the inherent polling and waiting states involved in collision avoidance for multiple access field-network systems. The best approach to eliminate time-delay is to avoid it altogether, through careful and proper system and industrial engineering practices. In the design of processes, careful considerations must be given to placement of sensors and the control devices to minimize dead times. However, under physical, spatial and operational constraints in the practical implementation of these systems, the extent to which this is realizable is fairly limited. Therefore, control of time-delay systems represents one important class of problems which has and will continue to pose a strong challenge for process control engineers.

1.2 Control of Time-Delay Systems

Owing to the time-delay present in delay systems, control action or the effect of control action cannot be realized until some time in the future when the system state variables may have changed considerably so that the control action is no longer appropriate. If a time-delay process is controlled by a PID controller, the derivative D-part of the controller is always turned off. To appreciate this conventional practice, assume that a set-point change is just being made and the PID controller is working to bring the process variable to the desired value. During the interval of the process dead time, the process variable has yet to react to the set-point change and therefore no dynamical information is available for the D-part to make the correct prediction. Therefore, if it is being used, the D-part will result in unnecessary oscillations in the system. On the other hand, with the D-part removed, the immediate disadvantage is no prediction from the controller on future control errors which is necessary for time-delay systems.

Therefore, it is important before control is attempted, that the system is well understood, in the sense of being modelled. Control systems are then designed so that the nominal system (model and controller) exhibits desirable behaviour. In what follows, two model-based predictive control schemes for time-delay processes are presented.

1.2.1 Smith-Predictor Control

The Smith-Predictor is probably the most popular strategy for control of time-delay systems. It was proposed by Smith (1957). Assume that a model for the process $g_p(s)$ is available which is described by

$$\hat{G}(s) = \hat{G}_0(s)e^{-sL},$$

where $\hat{G}_0(s)$ is a delay-free rational function and $\hat{L}$ is the time-delay.

The structure of the Smith-Predictor Controller is shown in Figure 1.1. It can be shown that the closed-loop transfer function is given by

$$G_{yr}(s) = \frac{G_c(s)G(s)}{1 + G_c(s)(\hat{G}_0(s) - \hat{G}(s) + G(s))}. \tag{1.1}$$

In the case of perfect modelling, i.e. $\hat{G}(s) = G(s)$, the closed-loop transfer function between the set-point and output is given by

$$G_{yr}(s) = G_{yr}^*(s) = \frac{G_c(s)\hat{G}_0(s)}{1 + G_c(s)\hat{G}_0(s)}e^{-\hat{L}s}. \tag{1.2}$$

This implies that the characteristic equation is free of the delay so that the primary controller $G_c(s)$ can be designed with respect to $\hat{G}_0(s)$. The achievable performance can thus be greatly improved over a conventional system without the delay-free output prediction.

One notable and not so desirable feature of the Smith-Predictor, however, is that it always retains the poles of the process. This has been shown by Watanabe and Ito (1981), Furukawa and Shimemura (1983) and Palmor and Ha levi (1983). Practical implications of the feature are limitations in the scope of application. The Smith-Predictor cannot be applied to unstable and poorly damped systems because the closed-loop cannot be stabilized or sufficiently stabilized.

1.2.2 Finite Spectrum Assignment

The Finite Spectrum Assignment (FSA) is an alternative time-delay compensation scheme. Unlike the Smith-Predictor, however, it can arbitrarily assign the closed-loop poles and therefore can be applied to poorly damped and unstable processes. Like the Smith-Predictor, the structure of the FSA is physically realizable and implementation can be as simple and efficient. One weakness of the FSA, as compared with Smith-Predictor, is probably that many developments in the FSA were only reported in recent years and the accessibility in terms of books and literature is not as extensive as the Smith-Predictor. Therefore, control engineers may not be as familiar and comfortable with the concept of the FSA. This book thus attempts to provide an extensive and self-sufficient treatment of the FSA, ranging from theoretical aspects of the controller to practical implementation issues.

1.3 Practical Issues

1.3.1 Automatic Tuning

An important feature of modern process control systems is the automatic tuning facility. With this feature, control engineers are spared the agony of manually tuning control loops which for large-scale processes, such as those in chemical processing and power plants, may be in the range of thousands in number. In particular, automatic tuning of PID controllers have been actively addressed in recent years. All methodologies are focused on setting up a simple system identification experiment effectively on an automated platform and using the process response generated from the experiment to tune the controller. One important development is accredited to the team of Astrom (1982) who pioneered the relay feedback system in auto-tuning of PID controllers. The

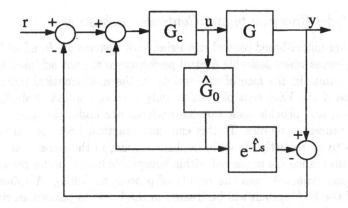

FIGURE 1.1. The Smith-Predictor Controller.

method has been extensively field tested and implemented in industrial controllers (e.g. the Satt Control Instruments) and found to be extremely useful and appropriate for its purpose.

It is useful to incorporate an auto-tuning facility in FSA systems. To this end, the book will present a possible extension of this important development to delay-systems and show how the FSA controller design may be realized following such an experiment.

1.3.2 Continuous Self-Tuning

Another trend in modern control systems is the incorporation of adaptive techniques and intelligence so that the systems are able to exploit available process information to react to changes in the same manner as a control engineer would have done under the same situation, albeit in a more efficient manner. Subsequent re-tuning of control systems will always be needed owing to time-varying drift of process characteristics and it is highly desirable if the control could take on the task independently. It will be therefore appropriate, in the book, to bundle state-of-the-art self-tuning and adaptation techniques available in the development and realization of FSA systems.

In certain cases, deliberate injection of excitation signals under normal mode of operation of process is not possible, reasons being that it is too dangerous in terms of possible human hazards posed, or too expensive in terms of possible economic loss owing to destruction and rejection of product batches. To this end, naturally occuring transients in the system such as those due to set-point changes or load disturbances can often reveal useful and sufficient updates on process dynamics to adapt the controller to changes in the process. A self-tuning method is introduced in the book, using practical *Discrete Fourier Transform* tools on the transients to estimate the frequency response at discrete frequencies. New controller parameters can then be realized based on transfer function models fitted to the frequency response.

1.3.3 Model Uncertainty and Controller Robustness

As with other model-based control strategies, FSA systems are faced with sensitivity problems when desirable control performance are not achieved within acceptable limits in the face of uncertainty in the mathematical models. It is of no use if the FSA control systems only achieve nominal stability and performance, and provide unacceptable performance under the slightest deviation in process dynamics. To this end, any practical FSA control system thus needs to be robustly viable in implementation, in the sense that stability and performance are preserved within acceptable limits, under reasonable perturbations expected from the results of process modelling. A robustness analysis of the FSA system will be treated in the book. In particular, robustness conditions will be derived including those for practical stability, robust stability and performance robustness which becomes guidelines for design of the controller. A detailed design methodology for time-delay systems with a first-order dynamical lag will be provided.

1.4 Real-Time Experimental Setup

Almost all results contained in this book have been tested in intensive simulation study as well as in real-time experiments. Two main experimental setups were used and they are briefly described in the following subsections.

1.4.1 Process Simulator

The *Dual process Simulator KI 100* from *KentRidge Instruments* is an analog process simulator which can be configured to simulate a wide range of industrial processes with different kinds of dynamics and at different levels of noise. The Simulator is connected to a PC via an A/D and D/A board. The windows-based *DT VEE 3.0* (Data Translation, 1995) is used as the system control platform, on which the control code is written in C++. The fastest sampling time of the *VEE* system is 0.06 sec.

1.4.2 Coupled-Tanks System

The experimental set-up of the coupled-tanks system is shown in Figure 1.2.

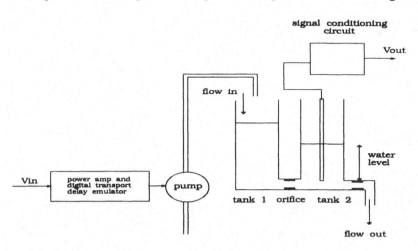

FIGURE 1.2. Schematic of the coupled-tanks system.

The pilot scale process consists of two square tanks, Tank 1 and Tank 2, coupled to each other through an orifice at the bottom of the tank wall. The inflow (control input) is supplied by a variable speed pump which pumps water from a reservoir into Tank 1 though a long tube. The orifice between Tank 1 and

Tank 2 allows the water to flow into Tank 2. In the experiments, it is desired to control the process with the voltage to drive the pump as input, and the water level in Tank 2 as process output. This coupled-tanks pilot process has process dynamics that are representative of many fluid level control problems faced in the process control industry.

FSA Algorithms Development

This chapter contains the control algorithms for Finite Spectrum Assignment (FSA). The development of four algorithms corresponding to varying degrees of system complexity are presented. The first, which is the basic one due to Ichikawa (1985), is derived under the assumption that the process contains distinct poles. The second algorithm relaxes this assumption and extends the basic FSA algorithm to systems with multiple poles. Both algorithms deal with single-variable processes. The FSA treatment on multivariable processes is provided in the third algorithm. Finally, the last algorithm provides asymptotic tracking and regulation of stable FSA systems operating under modelling errors.

2.1   Basic FSA Algorithm

The FSA algorithm for delay systems due to Ichikawa (1985) is actually an extension of Wolovich's frequency domain pole assignment for delay-free systems (Wolovich, 1974). It is thus helpful to briefly review the latter first.

Let a delay-free process be

$$Y(s) = G(s)U(s) = \frac{a(s)}{b(s)}U(s), \tag{2.1}$$

where $a(s)$ and $b(s)$ are polynomials of degree m and n respectively, $0 \leq m \leq n - 1$, and $a(s)$ and $b(s)$ are assumed to be co-prime. The process may be unstable and/or of non-minimum phase. Pole assignment is a means of stabilizing an unstable process.

Denote the n-degree monic polynomial desired as the characteristic polynomial of the closed-loop system by $p(s)$. Denote $b(s) - p(s)$ by $f(s)$, which is of degree $n - 1$ at most. Introduce any $(n - 1)$-degree monic asymptotically stable polynomial $q(s)$ (in what follows the term asymptotic will be omitted). The physical meaning of $q(s)$ is that it is the characteristic polynomial of a reduced-order Luenberger observer that yields a state estimate $\hat{x}(t)$ from the available $u(t)$ and $y(t)$, but the observer is constructed implicitly in the frequency pole assignment. Consider a polynomial equation:

$$k(s)b(s) + h(s)a(s) = q(s)f(s), \tag{2.2}$$

where $k(s)$ and $h(s)$ are unknown polynomials. Equation (2.2) yields a unique solution for $k(s)$ and $h(s)$ of degree at most $n - 2$ or $n - 1$ respectively. Using the solution $k(s)$ and $h(s)$, we construct the following control law:

$$U(s) = \frac{k(s)}{q(s)}U(s) + \frac{h(s)}{q(s)}Y(s) + R(s), \tag{2.3}$$

where $r(t)$ is an external reference input. The control law (2.3) achieves the desired pole assignment of $p(s)q(s)$ since (2.1) - (2.3) imply that

$$p(s)q(s)U(s) = b(s)q(s)R(s).$$

The control law (2.3) can be written in the time domain as

$$q(\mathbf{D})u(t) = k(\mathbf{D})u(t) + h(\mathbf{D})y(t) + q(\mathbf{D})r(t), \tag{2.4}$$

where $\mathbf{D}$ is the differential operator, *i.e.* $\mathbf{D}f(t) = \frac{df(t)}{dt}$, and the resulting control system is shown in Fig 2.1, where kq^{-1} is strictly proper while hq^{-1} is proper, and both are stable.

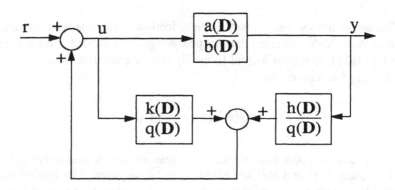

FIGURE 2.1. Pole-assignment control system for undelayed process.

Consider now a time-delay process:

$$Y(s) = G(s)U(s) = \frac{a(s)}{b(s)}e^{-Ls}U(s), \tag{2.5}$$

where $a(s)/b(s)$ is a strictly proper and coprime rational function of order n with $b(s)$ *monic*. $L > 0$ is a dead time, $b(s)$ is assumed to have no multiple zeros. The process with delay is of infinite dimension, and the whole state of the process is considered as the set of $x(t)$, the state of the lumped portion, and the time function $u(\tau)$, $t - L \leq \tau \leq t$. Naturally, the pole assignment will require the feedback of the whole state with appropriate coefficients. The state-space approach developed by Furukawa and Shimemura (1983) used another idea to feedback the predicted state $x(t + L)$. Since the predicted state, however, consists of $x(t)$ and $u(\tau)$, $t - L \leq \tau \leq t$, both ideas are considered to be equivalent.

The approach presented in this book depends rather on the former idea. The process state $x(t)$ can be estimated from $u(t - L)$ and $y(t)$, so that the first term in the control law (2.4) is changed to $k(\mathbf{D})u(t - L)$. In order to obtain the same result as was obtained by the time-domain method, some trial and error was needed in the determination of appropriate feedback coefficients for $u(\tau)$. Because $u(\tau)$ is infinite-dimensional, coefficients assume the form of a time function. Let the required closed-loop *monic* polynomial of degree n and the observer polynomial of degree $(n-1)$, respectively, be $p(s)$ and $q(s)$. Define

$$f(s) := b(s) - p(s), \qquad (2.6)$$

and consider the fraction $f(s)p^{-1}(s)$. Since $f(s)$ is of degree $n - 1$ at most, $p(s)$ is of degree n, and $p(s)$ is assumed to have no multiple zeros, then the partial fraction expansion of $f(s)p^{-1}(s)$ gives

$$\frac{f(s)}{b(s)} = \sum_{i=1}^{n} \frac{c_i}{s - \lambda_i}, \qquad (2.7)$$

where $\lambda_i, i = 1, 2, \cdots, n$, are n distinct poles of the process.

Define another polynomial $f_L(s)$ by

$$f_L(s) := b(s) \sum_{i=1}^{n} \frac{c_i e^{\lambda_i L}}{s - \lambda_i}, \qquad (2.8)$$

and the degree of $f_L(s)$ is at most $(n-1)$. One solves the following polynomial equation:

$$k(s)b(s) + h(s)a(s) = q(s)f_L(s), \qquad (2.9)$$

for $k(s)$ and $h(s)$ such that $k(s)/q(s)$ and $h(s)/q(s)$ are both proper. The control $u(t)$ is obtained from

$$q(\mathbf{D})u(t) = k(\mathbf{D})u(t - L) + h(\mathbf{D})y(t)$$
$$+ q(\mathbf{D}) \int_{-L}^{o} \sum_{i=1}^{n} c_i e^{-\lambda_i \tau} u(t + \tau)d\tau + q(\mathbf{D})r(t), \qquad (2.10)$$

where r is the set-point.

Theorem 2.1   *The control law (2.10) is realizable and achieves arbitrary finite spectrum assignment.*

PROOF   Since the polynomials $k(s)$ and $h(s)$ are of degree $n-2$ and $n-1$ (respectively) at most, both $k(s)q^{-1}(s)$ and $h(s)q^{-1}(s)$ are realizable. Furthermore, since $u(t+\tau)$, $-L \leq \tau \leq 0$, is the past history of the control signal over the finite interval L, and $c_i e^{-\lambda_i \tau}$ is definite over the interval, the integral $\int_{-L}^{0} c_i e^{-\lambda_i \tau} u(t+\tau)d\tau$ can be determined for any t. Therefore, the control law (2.10) is realizable in theory, although computation of the integral in practice requires some time.

Taking the Laplace transform of (2.10) yields

$$q(s)U(s) = k(s)U(s)e^{-Ls} + h(s)Y(s)$$
$$+q(s)\int_{-L}^{0}\sum_{i=1}^{n}c_i e^{-\lambda_i \tau}U(s)e^{\tau s}d\tau + q(s)R(s).$$

However, one notes that

$$\int_{-L}^{0}\sum_{i=1}^{n}c_i e^{-\lambda_i \tau}U(s)e^{\tau s}d\tau = \sum_{i=1}^{n}c_i\int_{-L}^{0}e^{(s-\lambda_i)\tau}d\tau U(s)$$
$$= \sum_{i=1}^{n}\frac{c_i}{s-\lambda_i}U(s) - \sum_{i=1}^{n}\frac{c_i}{s-\lambda_i}e^{\lambda_i L}U(s)e^{-Ls}$$
$$= \frac{f(s)}{b(s)}U(s) - \frac{f_L(s)}{b(s)}U(s)e^{-Ls}.$$

Therefore, we have

$$q(s)U(s) = k(s)U(s)e^{-Ls} + h(s)Y(s) + \frac{q(s)f(s)}{b(s)}U(s) \qquad (2.11)$$
$$-\frac{q(s)f_L(s)}{b(s)}U(s)e^{-Ls} + q(s)R(s).$$

On the other hand, multiplying $b^{-1}(s)U(s)e^{-Ls}$ to both sides of (2.9) yields

$$k(s)U(s)e^{-Ls} + h(s)Y(s) = \frac{q(s)f_L(s)}{b(s)}U(s)e^{-Ls}, \qquad (2.12)$$

It then follows from (2.11) and (2.12) that

$$q(s)U(s) = \frac{q(s)f(s)}{b(s)}U(s) + q(s)R(s),$$

or

$$q(s)p(s)U(s) = b(s)q(s)R(s),$$

So that the closed-loop poles are determined by $p(s)$ and $q(s)$ and they can be assigned arbitrarily.

One also sees that

$$Y(s) = \frac{a(s)}{b(s)}e^{-Ls}U(s) = \frac{q(s)a(s)}{q(s)p(s)}e^{-Ls}R(s),$$

However, since $q(s)$ is stable, the above reduces to

$$Y(s) = \frac{a(s)}{p(s)}e^{-Ls}R(s). \tag{2.13}$$

The FSA scheme is shown in Figure 2.2. It is interesting to note that although (2.11) is derived from (2.10), the control law must not be represented in the form of (2.11), because the third and fourth terms of (2.11) bring into the control system unnecessary dynamics, making the closed system unstable when the process is unstable.

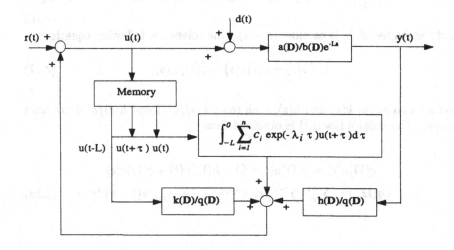

FIGURE 2.2. The FSA system

2.2   Processes with Multiple Poles

The basic FSA algorithm presented in Section 2.1 is developed under the assumption that the process has no multiple poles. The assumption will be removed in this section and a general frequency-domain finite spectrum assignment algorithm for delay processes with possibly multiple poles will be derived.

Consider again a delay system described by

$$Y(s) = G(s)U(s) = \frac{a(s)}{b(s)}e^{-Ls}U(s), \qquad (2.14)$$

where $a(s)/b(s)$ is a rational function of order n. Again, let the required closed-loop and observer polynomials, respectively, be $p(s)$ and $q(s)$. Define $f(s)$ same as in (2.6) where $b(s)$ and $p(s)$ have been assumed to be both monic, so that $f(s)$ is of degree $(n-1)$ at most. By partial fraction expansion, it follows that

$$\frac{f(s)}{b(s)} = \sum_{i=1}^{m} \sum_{j=1}^{v_i} \frac{c_{ij}}{(s - \lambda_i)^j}, \qquad (2.15)$$

where λ_i, $i = 1, 2, \cdots, m$, are m distinct zeros of $b(s)$, v_i are multiplicities of λ_i, and $\sum_{i=1}^{m} v_i = n$. Define another polynomial $f_L(s)$ by

$$f_L(s) := b(s) \sum_{i=1}^{m} \sum_{j=1}^{v_i} c_{ij} e^{\lambda_i L} \left[\frac{L^{j-1}}{(j-1)!(s-\lambda_i)} + \frac{L^{j-2}}{(j-2)!(s-\lambda_i)^2} + \cdots + \frac{1}{(s-\lambda_i)^j} \right],$$
$$(2.16)$$

and the degree of f_L is at most $(n-1)$. One solves the following equation:

$$k(s)b(s) + h(s)a(s) = q(s)f_L(s), \qquad (2.17)$$

to find a solution, $k(s)$ and $h(s)$, such that $k(s)/q(s)$ and $h(s)/q(s)$ are both proper. The control law $u(t)$ is obtained from

$$q(\mathbf{D})u(t) = k(\mathbf{D})u(t - L) + h(\mathbf{D})y(t) + q(\mathbf{D})r(t)$$
$$+ q(\mathbf{D}) \int_{-L}^{0} \sum_{i=1}^{m} \sum_{j=1}^{v_i} c_{ij} e^{-\lambda_i \tau} \frac{\tau^{j-1}}{(j-1)!(-1)^{j-1}} u(t + \tau) d\tau. \qquad (2.18)$$

Theorem 2.2   *The control law (2.18) is realizable and achieves arbitrary finite spectrum assignment.*

PROOF   The realizability of (2.18) can be shown in the same way as in the proof of Theorem 2.1. In what follows, it will be shown that the control law (2.18) assigns a finite spectrum characterized by $p(s)q(s)$ to the closed-loop system. Applying the shift theorem of the Laplace transform and the integral formula:

$$\int \tau^j e^{x\tau} d\tau = e^{x\tau} \left[\frac{\tau^j}{x} - j\frac{\tau^{j-1}}{x^2} + j(j-1)\frac{\tau^{j-2}}{x^3} - \cdots + (-1)^j j! \frac{1}{x^{j+1}} \right] + c,$$

where j is an positive integer, yields

$$L\left\{\int_{-L}^{0}\sum_{i=1}^{m}\sum_{j=1}^{v_i}c_{ij}e^{-\lambda_i\tau}\frac{\tau^{j-1}}{(j-1)!(-1)^{j-1}}u(t+\tau)d\tau\right\}$$

$$=\int_{-L}^{0}\sum_{i=1}^{m}\sum_{j=1}^{v_i}c_{ij}e^{(s-\lambda_i)\tau}\frac{\tau^{j-1}}{(j-1)!(-1)^{j-1}}U(s)d\tau$$

$$=U(s)\sum_{i=1}^{m}\sum_{j=1}^{v_i}\frac{c_{ij}}{(j-1)!(-1)^{j-1}}\int_{-L}^{0}\tau^{j-1}e^{(s-\lambda_i)\tau}d\tau$$

$$=U(s)\left\{\sum_{i=1}^{m}\sum_{j=1}^{v_i}\frac{c_{ij}}{(s-\lambda_i)^j}-e^{-Ls}\sum_{i=1}^{m}\sum_{j=1}^{v_i}\frac{c_{ij}}{(j-1)!}e^{\lambda_iL}\left[\frac{L^{j-1}}{s-\lambda_i}\right.\right.$$

$$\left.\left.+(j-1)\frac{L^{j-2}}{(s-\lambda_i)^2}+(j-1)(j-2)\frac{L^{j-3}}{(s-\lambda_i)^3}+\cdots+(j-1)!\frac{1}{(s-\lambda_i)^j}\right]\right\}$$

$$=\left[\frac{f(s)}{b(s)}-\frac{f_L(s)}{b(s)}e^{-Ls}\right]U(s).$$

Therefore, (2.18) can be expressed in the Laplace domain as

$$q(s)U(s)=\quad k(s)e^{-Ls}U(s)+h(s)Y(s)+\frac{q(s)f(s)}{b(s)}U(s)$$
$$-\frac{q(s)f_L(s)}{b(s)}e^{-Ls}U(s)+q(s)R(s). \qquad (2.19)$$

Substituting (2.14) into (2.19) gives the closed-loop system:

$$[q(s)b(s)-q(s)f(s)]Y(s)=\quad [k(s)b(s)+h(s)a(s)-q(s)f_L(s)]e^{-Ls}Y(s)$$
$$+a(s)q(s)e^{-Ls}R(s).$$

It follows from (2.6) and (2.17) that

$$q(s)p(s)Y(s)=a(s)q(s)e^{-Ls}R(s). \qquad (2.20)$$

Therefore, the closed-loop system has the characteristic polynomial $q(s)p(s)$ and this also means that the finite spectrum assignment has been achieved. The proof is completed.

The closed-loop transfer function between the output $Y(s)$ and the reference $R(s)$ is

$$G_{yr}(s)=\frac{a(s)}{p(s)}e^{-Ls}. \qquad (2.21)$$

Remark 2.1   When the process has no multiple poles, that is, $m=n, v_i=1, i=1,2,\cdots,n$, the control law in (2.18) reduces to the same formula as given by (2.10).

Remark 2.2   It should be noted that the integral in (2.18) defines a "static" mapping from the function space to the real space and causes no zero-pole cancellations. Thus, the method presented here can be applied to unstable processes, too.

Example 2.1 Consider a process described by

$$G(s) = \frac{s+1}{(s-1)^2}e^{-2s}.$$

Suppose that the closed-loop and observer polynomials are respectively

$$p(s) = (s+1)(s+2),$$

and

$$q(s) = (s+3).$$

It follows from the definitions before that

$$f(s) = b(s) - p(s) = -5s - 1,$$

$$\frac{f(s)}{b(s)} = \frac{-5s-1}{(s-1)^2} = -\frac{5}{s-1} - \frac{6}{(s-1)^2},$$

and

$$f_L(s) = (s-1)^2 \left\{ -5e^2 \left[\frac{1}{s-1} \right] - 6e^2 \left[\frac{2}{s-1} + \frac{1}{(s-1)^2} \right] \right\}$$
$$= e^2(11 - 17s).$$

Setting

$$k(s) = k,$$

and

$$h(s) = h_1 s + h_2,$$

and substituting them into (2.17) yield the system of linear equations:

$$\begin{bmatrix} 1 & 1 & 0 \\ -2 & 1 & 1 \\ 1 & 0 & 1 \end{bmatrix} \begin{bmatrix} k \\ h_1 \\ h_2 \end{bmatrix} = \begin{bmatrix} -17e^2 \\ -40e^2 \\ 33e^2 \end{bmatrix}.$$

The solution is

$$\begin{bmatrix} k \\ h_1 \\ h_2 \end{bmatrix} = \begin{bmatrix} 103.4468 \\ -229.0607 \\ 140.3921 \end{bmatrix},$$

and the required control law which achieves the assignment of the finite spectrum $\{-1, -2, -3\}$ is given in

$$(D+3)u(t) = 103.4468u(t-2) + (-229.0607D + 140.3921)y(t)$$
$$+ (D+3)r(t) + (D+3)\int_{-2}^{0} [-5+6\tau]e^{-\tau}u(t+\tau)d\tau.$$

2.3   Multivariable Processes

Modern industrial processes are usually complex and of a large-scale. They have multiple inputs and multiple outputs. The multivariable interaction between different loops has to be addressed in control design unless it is small enough to be neglected. Just like the single-variable case, the presence of considerable time delays in multivariable process is also well recognized, which makes the effective control of such processes even more difficult. A multivariable FSA may be a possible solution to this problem. In this section, the basic FSA in Section 2.1 is extended to linear time-invariant multivariable systems with multiple time delays.

Consider a multivariable system with multiple delays described by

$$Y(s) = G(s)U(s), \tag{2.22}$$

where Y and U are p-dimensional and m-dimensional vectors, respectively, and

$$G = [G_{ij}], \quad i = 1, 2, \ldots, p; \ j = 1, 2, \ldots, m;$$
$$G_{ij} = \sum_{k=1}^{\mu_{ij}} G_{ijk} \cdot e^{-L_{ijk}s}.$$

For simplicity in presentation, it is assumed that for any i, j and k, G_{ijk} are strictly proper rational functions with real coefficients which has no multiple poles.

The G_{ijk} have the following partial fraction forms:

$$G_{ijk} = \sum_{l=1}^{\nu_{ijk}} \frac{1}{s - \lambda_{ijkl}} \cdot \alpha_{ijkl}.$$

From a given G, we define a matrix G^0 with the same dimension as G:

$$\left.\begin{aligned}
G^0 &= [G_{ij}^0], \\
G_{ij}^0 &= \sum_{k=1}^{\mu_{ij}} G_{ijk}^0, \\
G_{ijk}^0 &= \sum_{l=1}^{\nu_{ijk}} \frac{1}{s - \lambda_{ijkl}} \cdot \beta_{ijkl}, \\
\beta_{ijkl} &= \alpha_{ijkl} e^{-L_{ijk}\lambda_{ijkl}},
\end{aligned}\right\} \tag{2.23}$$

where $i = 1, 2, \ldots, p; \ j = 1, 2, \ldots, m$. It is clear that G^0 has no delay at all and is a (strictly) proper rational function matrix. It follows that G^0 can be factorized as a right coprime polynomial matrix fraction:

$$G^0 = A(s)B^{-1}(s). \tag{2.24}$$

It is always possible (Wolovich 1974) to determine a polynomial matrix $P(s)$ which has column degrees equal to those of $B(s)$ such that $[\det P(s)]^{-1}$ has a preassigned set of closed-loop poles to be achieved by an appropriate control law, and F defined by

$$F = B - P, \tag{2.25}$$

has column degrees strictly less than those of B.

For the purpose of finite spectrum assignment, the following control law is introduced

$$Q(\mathbf{D})U(t) = K(\mathbf{D})U(t) + H(\mathbf{D})\{Y(t) + W[U(t)]\} + Q(\mathbf{D})R(t), \tag{2.26}$$

where $\mathbf{D}$ is the differential operator with respect to time t, R is a reference vector and Q, K and H are polynomial matrices which satisfy

$$K(s)B(s) + H(s)A(s) = Q(s)F(s), \tag{2.27}$$

with $[\det Q(s)]^{-1}$ being stable, and $Q^{-1}K$ and $Q^{-1}H$ proper (Wolovich 1974), and where

$$W[U(t)] = \begin{bmatrix} w_1 \\ w_2 \\ \vdots \\ w_p \end{bmatrix}, \tag{2.28}$$

$$w_i = \sum_{j=1}^{m} w_{ij},$$

$$w_{ij} = \sum_{k=1}^{\mu_{ij}} \int_0^{L_{ijk}} \left\{ \sum_{l=1}^{\nu_{ijk}} \beta_{ijkl} e^{\lambda_{ijkl} \cdot \tau} \right\} u_j(t - \tau) d\tau.$$

Now we are ready to establish the main result as follows.

Theorem 2.3   *The control law (2.26) achieves arbitrary finite spectrum assignment and is physically realizable.*

PROOF   Taking the Laplace transform of (2.26) yields

$$Q(s)U(s) = K(s)U(s) + H(s)[Y(s) + \mathbf{L}\{W[U(t)]\}] + Q(s)R(s). \tag{2.29}$$

In particular, one sees that

$$
\mathbf{L}\{w_{ij}\} = \sum_{k=1}^{\mu_{ij}} \int_0^{L_{ijk}} \left\{ \sum_{l=1}^{\nu_{ijk}} \beta_{ijkl} e^{\lambda_{ijkl}\tau} e^{-\tau s} \right\} d\tau \cdot u_j(s)
$$

$$
= \sum_{k=1}^{\mu_{ij}} \sum_{l=1}^{\nu_{ijk}} \frac{1}{s - \lambda_{ijkl}} \cdot \beta_{ijkl} \{ 1 - e^{L_{ijk}(\lambda_{ijkl} - s)} \} u_j(s)
$$

$$
= \sum_{k=1}^{\mu_{ij}} \sum_{l=1}^{\nu_{ijk}} \left[\frac{1}{s - \lambda_{ijkl}} \beta_{ijkl} - \frac{1}{s - \lambda_{ijkl}} \cdot \alpha_{ijkl} \cdot e^{-L_{ijk}s} \right] u_j(s).
$$

It follows from (2.22) and (2.23) that

$$
\begin{aligned}
\mathbf{L}\{w_{ij}\} &= (G_{ij}^0 - G_{ij})u_j(s), \\
\mathbf{L}\{W[U(t)]\} &= (G^0 - G)U(s) = G^0 U(s) - Y(s).
\end{aligned}
\tag{2.30}
$$

Substituting (2.30) into (2.29) gives

$$
Q(s)U(s) = K(s)U(s) + H(s)G^0(s)U(s) + Q(s)R(s). \tag{2.31}
$$

Equations (2.31) and (2.22) can be combined as

$$
\begin{bmatrix} Q - K - HG^0 & 0 \\ G & -I \end{bmatrix} \begin{bmatrix} U \\ Y \end{bmatrix} = \begin{bmatrix} Q \\ 0 \end{bmatrix} R.
$$

Then the spectrum of the closed-loop system is

$$
\sigma_s = \sigma \left(\det \begin{bmatrix} Q - K - HG^0 & 0 \\ G & -I \end{bmatrix} \right), \tag{2.32}
$$

where $\sigma(f)$ is the set of all complex numbers c such that $f(c) = 0$. By (2.24), (2.25) and (2.27), we obtain

$$
\sigma_s = \sigma[\det(Q)]\sigma[\det(P)]. \tag{2.33}
$$

This means that the closed-loop spectrum is finite and can be arbitrarily assigned by a suitable choice of Q and P. Note also that the integrals on the right-hand side of (2.26) require only the past history of control signals over some finite intervals, and $Q^{-1}K$ and $Q^{-1}H$ are proper. The control law (2.26) is therefore physically realizable. The theorem is thus true.

Remark 2.3   Poles λ_{ijkl} are not restricted to be real. In the case of complex conjugate poles, by some straightforward operations of complex numbers, it can be readily shown that G^0 and $W(U)$ still have real coefficients.

Example 2.2   The approach described above is now applied to the design of a basis weight and moisture content control system of a paper machine. The transfer function matrix relating basis weight and moisture content to the valve openings of thick pulp and steam flow is given by Wang (1984) as

$$
G = \begin{bmatrix} \dfrac{8.8}{s+0.5}e^{-s} & \dfrac{-0.81}{s+0.2}e^{-0.1s} \\[3mm] \dfrac{-0.75}{s+0.5}e^{-s} & \dfrac{-2.1}{s+0.2}e^{-0.1s} \end{bmatrix}.
$$

By the definition, we have

$$
G^0 = \begin{bmatrix} \dfrac{8.8}{s+0.5}e^{0.5} & \dfrac{-0.81}{s+0.2}e^{0.1\times0.2} \\[3mm] \dfrac{-0.75}{s+0.5}e^{0.5} & \dfrac{-2.1}{s+0.2}e^{0.1\times0.2} \end{bmatrix} = \begin{bmatrix} \dfrac{14.5}{s+0.5} & \dfrac{-0.826}{s+0.2} \\[3mm] \dfrac{-1.24}{s+0.5} & \dfrac{-2.14}{s+0.2} \end{bmatrix}.
$$

Its right coprime fraction is

$$
G^0 = AB^{-1} = \begin{bmatrix} 14.5 & -0.826 \\ -1.24 & -2.14 \end{bmatrix} \begin{bmatrix} s+0.5 & 0 \\ 0 & s+0.2 \end{bmatrix}^{-1}
$$

P and Q are chosen as

$$
P = \begin{bmatrix} s+2 & 0 \\ 0 & s+1 \end{bmatrix}, Q = \begin{bmatrix} s+10 & 0 \\ 0 & s+5 \end{bmatrix}, \tag{2.34}
$$

with the closed-loop poles being -1, -2, -5, -10. It then follows that

$$
F = B - P = \begin{bmatrix} -1.5 & 0 \\ 0 & -0.8 \end{bmatrix}.
$$

In order to solve the equation:

$$
KB + HA = QF, \tag{2.35}
$$

for K and H, B and Q are rewritten as

$$
B = B_1 s + B_0, \quad Q = Q_1 s + Q_0, \quad A = A_0, \quad F = F_0, \tag{2.36}
$$

where $B_1 = Q_1 = I_2$ and

$$
B_0 = \begin{bmatrix} 0.5 & 0 \\ 0 & 0.2 \end{bmatrix}, Q_0 = \begin{bmatrix} 10 & 0 \\ 0 & 5 \end{bmatrix}.
$$

Setting

$$H = I_2 s + H_0, \quad K = K_0 \tag{2.37}$$

and substituting (2.36) and (2.37) into (2.35) yield

$$K_0 + A_0 = F_0,$$
$$K_0 B_0 + H_0 A_0 = Q_0 F_0.$$

The solution is

$$K = K_0 = \begin{bmatrix} -16 & 0.826 \\ 1.24 & 1.34 \end{bmatrix}, \tag{2.38}$$

$$H_0 = \begin{bmatrix} -0.46 & 0.26 \\ 0.12 & 1.9 \end{bmatrix},$$

$$H = I_2 s + H_0 = \begin{bmatrix} s - 0.46 & 0.26 \\ 0.12 & s + 1.9 \end{bmatrix}. \tag{2.39}$$

$Q^{-1}K$ and $Q^{-1}H$ are obviously proper. Finally, the control law is

$$Q(\mathbf{D}) \begin{bmatrix} u_1(t) \\ u_2(t) \end{bmatrix} = K(\mathbf{D}) \begin{bmatrix} u_1(t) \\ u_2(t) \end{bmatrix} + Q(\mathbf{D})R(t)$$

$$+ H(\mathbf{D}) \times \left\{ \begin{bmatrix} y_1(t) \\ y_2(t) \end{bmatrix} + \begin{bmatrix} 14.5 & -0.826 \\ -1.24 & -2.14 \end{bmatrix} \times \begin{bmatrix} \int_0^1 e^{-0.5\tau} u_1(t-\tau) d\tau \\ \int_0^{0.1} e^{-0.2\tau} u_2(t-\tau) d\tau \end{bmatrix} \right\},$$

where Q, K and H are given in (2.34), (2.38) and (2.39), respectively. It follows from (2.31) that $U(s) = BP^{-1}R(s)$. With $Y = GU$, then the resulting closed-loop transfer matrix relating Y to R is

$$G_{YR} = GBP^{-1} = \begin{bmatrix} \dfrac{8.8}{s+2} e^{-s} & \dfrac{-0.81}{s+1} e^{-0.1s} \\ \dfrac{-0.75}{s+2} e^{-s} & \dfrac{-2.1}{s+1} e^{-0.1s} \end{bmatrix}.$$

Compared with G, it can be seen that all the demonstrators of elements of G have been changed into the desired ones while the rest remain unchanged.

2.4 Tracking and Regulation

Asymptotic tracking and regulation have always been desirable properties for any process control systems. It is, however, noted that a non-zero steady state error may exist even when an integrator is cascaded with the process and the FSA is applied to the cascaded system in the usual way.

Example 2.3   Consider the process:

$$G_p(s) = \frac{1}{5s+1}e^{-10s},$$

which is cascaded with an integrator so that the generalized process is

$$G(s) = \frac{1}{s(5s+1)}e^{-10s}.$$

To design a FSA for a delay system, a $p(s)$ may be chosen such that $p(0) = a(0)$ so that from (2.13) the closed-loop transfer function:

$$G_{yr}(s) = \frac{a(s)}{p(s)}e^{-Ls},$$

has a static gain of 1 to achieve asymptotic tracking. According to (2.5), $a(s) = \frac{1}{5}$ and $a(0) = 0.2$, $p(s)$ is then specified as

$$p(s) = s^2 + 0.894s + 0.2,$$

with the damping factor of 1, and the resultant closed-loop transfer function

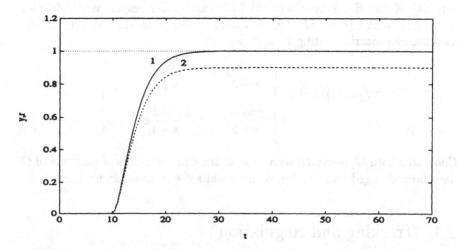

FIGURE 2.3. Performance of the FSA system for (1) the original process , (2) the process after a change in parameter.

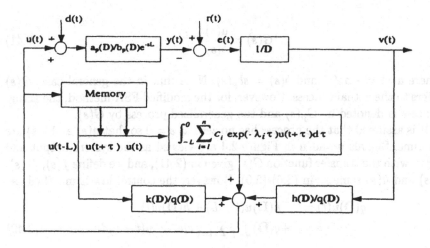

FIGURE 2.4. The modified FSA system.

is

$$G_{yr}(s) = \frac{0.2}{s^2 + 0.894s + 0.2} e^{-10s}.$$

It has no steady-state error in response to a step set-point change. However, the static gain cannot be guaranteed to be 1 if there is a perturbation in the process. For instant, if $a(s)$ is changed from 0.2 to 0.18, the transfer function becomes

$$\tilde{G}_{yr}(s) = \frac{0.18}{s^2 + 0.894s + 0.2} e^{-10s}.$$

The closed-loop system then has a steady state error. This is shown in Figure 2.3. Therefore, the incorporation of an integrator into the FSA systems usually cannot remove offset because of inevitable modelling errors.

In order to achieve asymptotic tracking and regulation in presence of possible modelling errors, a modified FSA scheme is proposed and shown in Figure 2.4. Basically, the error between the set-point and process output is formed after the process and this is followed by an integrator. The resultant generalized process is then stabilized with the ordinary FSA method.

Let a process be described by

$$Y(s) = G_p(s)U(s) = \frac{a_p(s)}{b_p(s)} e^{-Ls} U(s),   (2.40)$$

where $a_p(s)/b_p(s)$ is a proper and coprime rational function with $b_p(s)$ monic. The generalized process consisting of the process and an integrator is represented by the transfer function:

$$G(s) = \frac{a(s)}{b(s)} e^{-Ls}, \tag{2.41}$$

where $a(s) = -a_p(s)$ and $b(s) = sb_p(s)$. Note that in the general case, $G(s)$ refers to the actual process. However, for the modified FSA method, the actual process is denoted by $G_p(s)$ and the generalized process by $G(s)$.

It is assumed that the process has no zero at $s = 0$ so that $a(s)$ and $b(s)$ are coprime. Signals v and u in Figure 2.4 are viewed as the system output and input with the transfer function $G(s)$ given in (2.41), and we define $f(s)$, $f_L(s)$, $k(s)$ and $h(s)$ same as in (2.6)-(2.9). However, the control law is modified as

$$q(\mathbf{D})u(t) = \quad k(\mathbf{D})u(t-L) + h(\mathbf{D})v(t)$$
$$+q(\mathbf{D}) \int_{-L}^{o} \sum_{i=1}^{n} c_i e^{-\lambda_i \tau} u(t+\tau) d\tau \tag{2.42}$$

The differences between (2.10) and (2.42) are that the system output y in (2.10) is replaced with the generalized process output v in (2.42), and that the set-point r in (2.10) disappeared in (2.42) as it now enters the system through the generalized process. Taking the Laplace transform of (2.42) yields

$$q(s)U(s) = \quad k(s)e^{-Ls}U(s) + h(s)V(s) + \frac{q(s)f(s)}{b(s)}U(s)$$
$$-\frac{q(s)f_L(s)}{b(s)}e^{-Ls}U(s) \tag{2.43}$$

The closed-loop relationship is now given by

$$Y(s) = \frac{-h(s)a(s)}{p(s)q(s)} e^{-Ls} R(s), \tag{2.44}$$

and the nominal system is stable provided that $p(s)$ and $q(s)$ are Hurwitz polynomials.

Theorem 2.4   *The system in Figure 2.4 achieves asymptotic tracking and regulation in response to a step set-point change and/or load disturbance provided that it is stable.*

PROOF.   With (2.6), (2.9), (2.40), (2.42), it follows from Figure 2.4 that the error between the set-point and output is

$$E(s) := R(s) - Y(s) = G_{er}(s)R(s) + G_{ew}(s)W(s), \tag{2.45}$$

where

$$G_{er}(s) = \frac{q(s)p(s) + h(s)a(s)e^{-Ls}}{q(s)p(s)}, \tag{2.46}$$

and

$$G_{ew}(s) = -G_p(s)\frac{q(s)p(s) + h(s)a(s)e^{-Ls}}{q(s)p(s)}. \tag{2.47}$$

Define

$$\phi(s) = q(s)p(s) + h(s)a(s)e^{-Ls}. \tag{2.48}$$

It is now claimed that

$$\phi(\lambda_i) = 0, \quad i = 1, 2, \cdots, n, \tag{2.49}$$

that is, each zero λ_i of $b(s)$ is that of $\phi(s)$. Indeed, at $s = \lambda_i$, (2.9) becomes

$$h(\lambda_i)a(\lambda_i) = q(\lambda_i)f_L(\lambda_i).$$

It follows from (2.8) and (2.7) that

$$f_L(\lambda_i) = e^{\lambda_i L} \cdot c_i \frac{b(s)}{s - \lambda_i} |_{s=\lambda_i} = e^{\lambda_i L} f(\lambda_i).$$

Hence, we have

$$\phi(\lambda_i) = q(\lambda_i)p(\lambda_i) + q(\lambda_i)f(\lambda_i).$$

Furthermore, it follows from (2.6) that

$$\phi(\lambda_i) = q(\lambda_i)b(\lambda_i) = 0,$$

and thus (2.49) is true. For a step disturbance, $W(s) = \frac{\beta}{s}$, one sees from (2.47) and (2.48) that

$$G_{ew}(s)W(s) = \frac{\beta a(s)e^{-Ls}}{q(s)p(s)} \cdot \frac{\phi(s)}{b(s)}, \tag{2.50}$$

(2.49) implies that $\phi(s)$ cancels $b(s)$ completely and (2.50) is thus stable since both $q(s)$ and $p(s)$ are stable by design. Similarly, for a step set-point change, $R(s) = \frac{\alpha}{s}$, it follows from (2.46) that

$$G_{er}(s)R(s) = \frac{\alpha b_p(s)}{q(s)p(s)} \cdot \frac{\phi(s)}{b(s)}, \tag{2.51}$$

which is, again by (2.49), stable, too. The stability of (2.50) and (2.51) together implies that $E(s)$ in (2.45) is stable and $e(\infty) = 0$. Hence the result.

Remark 2.4   If a set-point change or load disturbance is of other types than step one, the integrator in Figure 2.4 should be replaced with $\frac{1}{\psi(s)}$, where $\psi(s)$ is the least common denominator of set-point and disturbance signals in the Laplace domain. The asymptotic tracking and regulation are still achievable with the generalized process, $\frac{-G_p(s)}{\psi(s)}$, stabilized with the FSA scheme.

Remark 2.5   If, with $\frac{1}{\psi(s)}$ absorbed into the process, the generalized process has a multiple pole, then the FSA algorithm developed in Section 2.2 should be used instead of the one described in Section 2.1 which is applicable only to systems with distinct poles. With this modification, it is straightforward to show that Theorem 2.1 remains true.

Remark 2.6   The modified FSA system proposed above will maintain the asymptotic tracking and regulation property in the presence of a process perturbation provided that the perturbed system remains stable. Physically, when the system stability is preserved under perturbations in the system, all signals in the Figure 2.4 will become finite constants as the time approaches infinity. But, in order for the output of the integrator, v, to maintain constant in the steady state, its input, e, must be zero, implying that asymptotic tracking and regulation have been achieved. Due to this useful property of the method, the modified FSA algorithm will be the main one of interest in the rest of the book.

Remark 2.7   If the given process is of multivariable, then in the modified FSA system in Figure 2.4, the block following the error $e(t)$ should have $diag\{\frac{1}{\psi(D)}\}$ to replace $(\frac{1}{D})$ and the FSA algorithm for multivariable system in Section 2.3 should be applied to $G(s) = diag\{\frac{1}{\psi(s)}\}(-G_p)$ to find a stabilizing and arbitrary finite spectrum assignment controller.

2.5   Problems

1. Consider a process described by the following transfer function :

$$G(s) = \frac{Y(s)}{U(s)} = \frac{1}{1-s}e^{-5s}.$$

Assume that the desired pole is at -1 for the closed-loop and at -5 for the observer. Compute the FSA control law to achieve it.

2. Consider a process described by the following transfer function :

$$G(s) = \frac{Y(s)}{U(s)} = \frac{1}{(1-s)^2}e^{-20s},$$

stabilize the process using FSA control.

3. Consider a process described by the following transfer function :

$$G(s) = \frac{Y(s)}{U(s)} = \frac{2}{5s+1}e^{-20s}.$$

Design an FSA control law to ensure asymptotic tracking and regulation of the process variable.

4. What are guidelines for selection of the observer poles relative to the underlying process dynamics?

5. An algorithm for multivariable dealy systems without multiple poles is presented in Section 2.3. Extend this algorithm to the multivariable systems with multiple poles using the result in Section 2.2.

6. For stable processes, compare the FSA with popular Smith Predictor Control scheme in terms of structure, performance, design and implementation complexity.

7. Write a simulation program to implement the basic algorithm in Section 2.1 (use of MATLAB/SIMULINK is recommended).

8. Can the FSA scheme be extended to nonlinear processes?

Closed-loop Process Identification

3.1 Introduction

A primary requirement for the realization of FSA systems is the availability of a mathematical model for the actual process, and the ways to obtain this model may be broadly categorized into 2 methodologies: physical modelling techniques or process identification methods. The physical modelling approach seeks to derive the process model using fundamental laws governing the operations of the process, such as continuity and energy equations. Therefore, the parameters in the model have physical interpretations and they may sometimes be obtained from direct measurements on the process. Clearly, this modelling method requires a good and thorough understanding of the functioning of the entire process. However, even when the requirement can be satisfied, rarely can all parameters of the model be obtainable through direct measurements. Therefore, often physical modelling methods cannot be used alone and they are complemented with other modelling techniques.

Process identification, on the other hand, is a more generally applicable method which seeks to obtain a mathematical description of the actual process based on the input and output signals. Excitation signals are injected into the process, which is treated as a black box, and the response from the process is logged for analysis to yield a mathematical relationship between the process input and output. This may be done either with the process in the open-loop or closed-loop under normal feedback operations, and analysis may be done in the time or frequency domain. In the time-domain, the more commonly used methods are transient analysis and correlation analysis to obtain the impulse response of the system which will ideally characterize the system. In the frequency domain, the common methods are frequency correlation, Fourier and spectral analysis to give the frequency response of the system. One of the most important phases in process identification is the experiment design phase. The decisions on certain design attributes of the experiment are crucial to the success of the modelling effort, such as the choice of input signals, the choice of sampling interval and the choice of an adequate model structure, and all these decisions must be closely-knitted to the characteristics of the process. In the absence of sufficient information on process characteristics, an effective choice of these design attributes has always been a very difficult task.

An interesting experiment design for process frequency response estimation which alleviates some of the experimental design problems is the relay feed-back system shown in Figure 3.1, first pioneered by Astrom and co-workers. This method has been the subject of much interest in recent years and it has been field tested in a wide range of applications. There are many attractive features associated with the relay feedback technique. First, for most industrial processes, the arrangement automatically results in a sustained oscillation approximately at the ultimate frequency of the process. From the oscillation amplitude, the ultimate gain of the process can be estimated. This alleviates the task of input specification from the user and therefore is in contrast to other frequency-domain based methods requiring the frequency characteristics of the

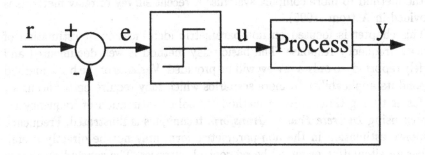

FIGURE 3.1. Relay feedback system.

input signal to be specified. The particular feature of the relay feedback technique greatly facilitates automatic tuning procedures since the arrangement will automatically give an important point of the process frequency response. Secondly, the relay feedback method is a closed-loop test and the process variable is maintained around the set-point value. This keeps the process in the linear region where the frequency response is of interest which works well on highly nonlinear processes ; the process is never pushed very far away from the steady-state conditions. Current identification techniques relying on transient analysis such as impulse or step tests do not possess this property, and they are therefore ineffective for processes with nonlinear dynamics. Thirdly, the relay feedback technique does not require prior information of the system time constants for a careful choice of the sampling period. The choice of the sampling period has always been a tricky problem for traditional parameter estimation techniques. If the sampling interval is too long, the dynamics of the process will not be adequately captured in the data and the accuracy of model subsequently obtained will be poor as a consequence. While a conservative safety-first approach towards this decision may be to select the smallest sampling period supported by the data acquisition equipment, this would result in too much data collection with inconsequential information. A corrective action then is data decimating in the post-treatment phase which for real-time parameter estimation may not be tolerable. With these features, the relay feedback method is therefore an attractive method to consider in auto-tuning applications.

Relay systems can be traced back to their classical configurations. In the fifties, relays were mainly used as amplifiers but such applications are obsolete now, owing to the development of electronic technology. In the sixties, relay feedback was applied to adaptive control. One prominent example of such applications is the self-oscillating adaptive controller developed by Minneapolis Honeywell which uses relay feedback to attain a desired amplitude margin. This system was tested extensively for flight control systems, and it has been used in several missiles. It was in the eighties that Astrom successfully applied the relay feedback method to auto-tune PID controllers for process control, and triggered a resurgence of interest in relay methods, including extensions of the method to more complex systems. A recent survey of relay methods is provided in Astrom (1995).

This chapter is focused on non-parametric identification or estimation of process frequency response. The basic relay method is well documented and widely reported, so only a review will be provided. Variants of the basic method expand its applicability to more scenarios which may require better accuracy or faster tuning time. Next, a method for online estimation of frequency response using *Discrete Fourier Transform* techniques is illustrated. Frequency response estimation in the non-parametric form may not be directly useful, either for simulation or model-based control purposes. The concluding section in the chapter elaborates how common transfer function models may be easily fitted to the frequency response obtained from the relay-based identification methods. The use of these methods for auto-tuning of the FSA system will be covered in Chapter 4.

3.2   Basic Relay

The ultimate frequency ω_π of a process, where the phase lag is $-\pi$, can be determined automatically from an experiment with relay feedback as shown in Figure 3.1. By making the observation that the describing function of a relay is the negative real axis, it follows that the system oscillates with a frequency that is close to ω_π. The process ultimate gain k_π is then approximately

$$k_\pi = \frac{4\mu}{\pi a},\tag{3.1}$$

where μ and a are the amplitudes of relay and process output respectively. Equation (3.1) follows from a describing function analysis (Subsection 3.4.1).

It may be advantageous to use a relay with hysteresis as shown in Figure 3.2 so that the resultant system is less sensitive to measurement noise. The inverse negative describing function of this relay is given by $-\frac{1}{N(a)} = -\frac{\pi}{4\mu}\left(\sqrt{a^2 - \epsilon^2} + j\epsilon\right)$. In this case, the oscillation corresponds to the point where the negative inverse describing function of the relay crosses the Nyquist curve of the process as shown in Figure 3.3. With hysteresis, there is an additional parameter ϵ which can, however, be set automatically based on a pre-determination of the measurement noise level.

Relay tuning is an attractively simple method for extracting the critical point of a process. Accompanying the method are two main limitations. First, the accuracy of the estimation could be poor for certain processes. Secondly, the relay experiment yields only one point of the process frequency response which may not be adequate in many applications other than raw PID tuning. To overcome these limitations, some modifications to the basic method are necessary and these are covered in the subsequent sections.

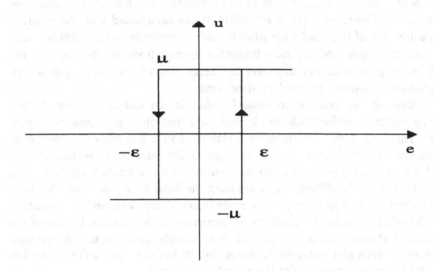

FIGURE 3.2. Unbiased Relay

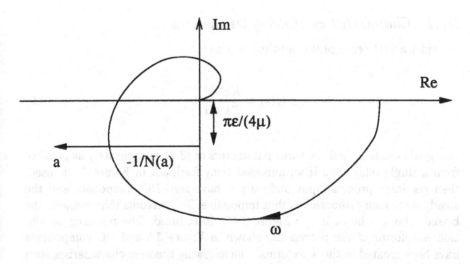

FIGURE 3.3. Negative inverse describing function of the hysteretic relay.

3.3 Relay with Bias

Model-based controllers need transfer function models (Hang *et al.* 1995; Pal-mor *et al.* 1994; Wang *et al.* 1995). The most commonly used dynamic model for industrial processes is the first-order plus dead-time model, which has three parameters. However, the ultimate gain and ultimate frequency obtained by us-ing Astrom's basic method are not sufficient to find a transfer function model. The additional information is required for this purpose. Luyben (1987) sug-gests that the dead time is read off from the initial response of the system to the relay test or the steady-state gain is obtained from a steady-state model of the process. An alternative is to run a second relay test with a dead time cascaded to the process (Li *et al.*, 1991). It has been noted that the accurate measurement of the dead time from the initial response is very difficult, and the ultimate gain and ultimate frequency derived from the describing func-tion are approximate and may have significant errors in case of processes with high-order dynamics and/or long dead time.

In this section, exact expressions for the periods and amplitudes of limit cycles under relay feedback are derived for a first-order plus dead time pro-cess. This time-domain information is combined with frequency response point estimation using Fourier series expansions of the limit cycles so that a first-order plus dead time model can be identified with a single relay test. This method avoids the difficulty of measuring the dead time from the relay test. Furthermore, no approximation is made in our derivations and the resultant model will be precise if it matches the structure of the process. In case of the mismatched structure, it is shown through extensive simulations that our pro-cedure is robust and yields very accurate results in the sense that the identified model frequency response fits the actual process well.

3.3.1 Characteristics of Relay Oscillations

Consider a first-order plus dead-time process:

$$G(s) = \frac{Ke^{-Ls}}{Ts+1}. \tag{3.2}$$

Our goal here is to find the three parameters in (3.2) as accurately as possible from a single relay test. If an unbiased relay feedback in Figure 3.2 is used, then resultant process input and output have zero DC component and the steady state gain estimation is thus impossible. To overcome this problem, the biased relay as shown in the Figure 3.4 is introduced. The resulting oscilla-tion waveforms of the process are shown in Figure 3.5 and DC components have been created in the waveforms. The following theorem characterises such oscillations.

Theorem 3.1   *For the process of (3.2) under the relay feedback of Figure 3.4, the process output y converges to the stationary oscillation in one period $(T_{u1} + T_{u2})$, and the oscillation is characterized by*

$$A_u = (\mu_0 + \mu)K(1 - e^{-\frac{L}{T}}) + \varepsilon e^{-\frac{L}{T}}, \qquad (3.3)$$

$$A_d = (\mu_0 - \mu)K(1 - e^{-\frac{L}{T}}) - \varepsilon e^{-\frac{L}{T}}, \qquad (3.4)$$

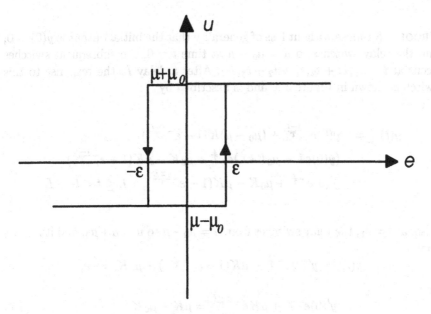

FIGURE 3.4. Biased Relay

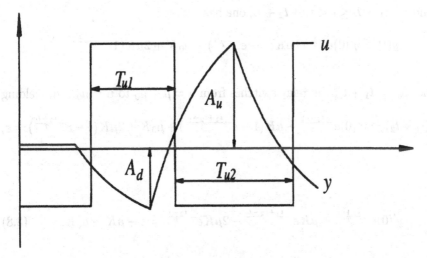

FIGURE 3.5. Oscillatory Waveforms under a Biased Relay Feedback

$$T_{u1} = T \ln \frac{2\mu K e^{\frac{L}{T}} + \mu_0 K - \mu K + \varepsilon}{\mu K + \mu_0 K - \varepsilon},$$
(3.5)

and

$$T_{u2} = T \ln \frac{2\mu K e^{\frac{L}{T}} - \mu_0 K - \mu K + \varepsilon}{\mu K - \mu_0 K - \varepsilon}.$$
(3.6)

PROOF: Suppose without loss of generality that the initial output is $y(0) > 0$, and the relay switches to $u = \mu_0 - \mu$ at time $t = 0$, the subsequent switches occur at $t = t_1, t_1 + t_2, t_1 + t_2 + t_3, \cdots$. After a delay L, the response to this switch is shown in Figure 3.5, and is described by

$$
\begin{aligned}
y(t) &= y(0)e^{-\frac{t-L}{T}} + (\mu_0 - \mu)K(1 - e^{-\frac{t-L}{T}}) \\
&= (y(0)e^{\frac{L}{T}} - \mu_0 K e^{\frac{L}{T}})e^{-\frac{t}{T}} + \mu_0 K - \mu K(1 - e^{-\frac{t-L}{T}}) \\
&:= y'(0)e^{-\frac{t}{T}} + \mu_0 K - \mu K(1 - e^{-\frac{t-L}{T}}), \quad L \leq t < t_1 + L.
\end{aligned}
$$

Also, at $t = t_1$, the relay switches from $u = \mu_0 - \mu$ to $u = \mu + \mu_0$, and it follows that

$$y(t_1) = y'(0)e^{-\frac{t_1}{T}} - \mu K(1 - e^{-\frac{t_1-L}{T}}) + \mu_0 K = -\varepsilon,$$

or

$$y'(0)e^{-\frac{t_1}{T}} + \mu K e^{-\frac{t_1-L}{T}} = \mu K - \mu_0 K - \varepsilon.$$
(3.7)

And for $t_1 + L \leq t < t_1 + t_2 + L$, one has

$$y(t) = y'(0)e^{-\frac{t}{T}} - \mu K(1 - e^{-\frac{t-L}{T}}) + \mu_0 K + 2\mu K(1 - e^{-\frac{t-t_1-L}{T}}),$$

and at $t = t_1 + t_2$, the relay switches from $u = \mu + \mu_0$ to $u = \mu_0 - \mu$, yielding

$$y(t_1 + t_2) = y'(0)e^{-\frac{t_1+t_2}{T}} - \mu K(1 - e^{-\frac{t_1+t_2-L}{T}}) + \mu_0 K + 2\mu K(1 - e^{-\frac{t_2-L}{T}}) = \varepsilon,$$

or

$$y'(0)e^{-\frac{t_1+t_2}{T}} + \mu K e^{-\frac{t_1+t_2-L}{T}} - 2\mu K e^{-\frac{t_2-L}{T}} = \varepsilon - \mu K - \mu_0 K.$$
(3.8)

Substituting (3.7) into (3.8) gives

$$(\mu K - \mu_0 K - \varepsilon)e^{-\frac{t_2}{T}} + (\mu + \mu_0)K - 2\mu K e^{-\frac{t_2 - L}{T}} = \varepsilon,$$

and

$$t_2 = T \ln \frac{2\mu K e^{\frac{L}{T}} + \mu_0 K - \mu K + \varepsilon}{\mu K + \mu_0 K - \varepsilon}.$$

Also, it is clear that

$$y(t) = y'(0)e^{-\frac{t}{T}} - \mu K(1 - e^{-\frac{t-L}{T}}) + 2\mu K(1 - e^{-\frac{t-t_1-L}{T}})$$
$$-2\mu K(1 - e^{-\frac{t-t_1-t_2-L}{T}}) + \mu_0 K, \quad t_1 + t_2 + L \le t < t_1 + t_2 + t_3 + L.$$

At $t = t_1 + t_2 + t_3$, the relay switches from $u = \mu_0 - \mu$ to $u = \mu + \mu_0$, leading to

$$y(t_1 + t_2 + t_3) = y'(0)e^{-\frac{t_1+t_2+t_3}{T}} - \mu K(1 - e^{-\frac{t_1+t_2+t_3-L}{T}})$$
$$+2\mu K(1 - e^{-\frac{t_2+t_3-L}{T}}) - 2\mu K(1 - e^{-\frac{t_3-L}{T}}) + \mu_0 K$$
$$= -\varepsilon.$$

Using (3.8) yields

$$(\varepsilon - \mu_0 K - \mu K)e^{-\frac{t_3}{T}} + (\mu_0 - \mu)K + 2\mu K e^{-\frac{t_3-L}{T}} = -\varepsilon,$$

and it follows that

$$t_3 = T \ln \frac{2\mu K e^{\frac{L}{T}} - \mu K - \mu_0 K + \varepsilon}{\mu K - \mu_0 K - \varepsilon}.$$

Similarly, one can show that

$$t_2 = t_4 = t_6 = \cdots := T_{u1},$$

and

$$t_3 = t_5 = t_7 = \cdots := T_{u2}.$$

The down-amplitude A_d and the up-amplitude A_u of the oscillation can be computed from the output $y(t)$ at $t = t_1 + L$ and $t = t_1 + t_2 + L$ respectively as

$$A_d = y(t_1 + L) = (\mu_0 - \mu)K(1 - e^{-\frac{L}{T}}) - \varepsilon e^{-\frac{L}{T}},$$

$$A_u = y(t_1 + t_2 + L) = (\mu_0 + \mu)K(1 - e^{-\frac{L}{T}}) + \varepsilon e^{-\frac{L}{T}},$$

which completes the proof.

Equations (3.3) - (3.6) are the exact expressions of the period and the amplitude of limit cycle oscillations for the process (3.2). The result of several simulation examples are given in Table 3.1, where the outputs of biased relay are 1.3 and -0.7 respectively, and the hysteresis of relay is 0.1. The results show the accuracy of the formulas (3.3) - (3.6), and very small errors are caused by simulation computations.

3.3.2   Transfer Function Modelling

The four equations (3.3) - (3.6) are sufficient to determine the three parameters of the process model (3.2), but solving these equations is somehow tedious. To simplify the computations, we now look into frequency response information contained in limit cycles. Theorem 3.1 indicates that the waveforms of the process input $u(t)$ and output $y(t)$ are periodic with the period $T_{u1} + T_{u2}$. They can be expanded into Fourier series. The direct-current components of these periodic waves are extracted, and the steady state gain of the process can be computed (Ramirez, 1985) via the following formula:

$$K = G(0) = \frac{\int_0^{T_{u1}+T_{u2}} y(t)dt}{\int_0^{T_{u1}+T_{u2}} u(t)dt}. \tag{3.9}$$

With K known, the normalized dead time of the process $\Theta = \frac{L}{T}$ is obtained from (3.3) or (3.4) as

$$\Theta = \ln \frac{(\mu_0 + \mu)K - \varepsilon}{(\mu_0 + \mu)K - A_u}, \tag{3.10}$$

Table 3.1. The Limit Cycle under the Biased Relay Feedback

Process			Calculated Result				Measured Result				Percentage Error %			
K	T	L	T_{u1}	T_{u2}	A_u	A_d	T_{u1}	T_{u2}	A_u	A_d	T_{u1}	T_{u2}	A_u	A_d
1	2	1	1.620	2.503	0.5722	-0.3361	1.620	2.505	0.572	-0.336	0.000	0.080	0.035	0.030
1	2	2	2.788	3.909	0.8585	-0.4793	2.790	3.910	0.859	-0.479	0.072	0.026	0.058	0.063
1	2	5	5.972	7.307	1.2015	-0.6507	5.970	7.310	1.202	-0.651	0.034	0.041	0.042	0.046
1	1	2	2.469	3.119	1.1376	-0.6188	2.470	3.120	1.138	-0.619	0.041	0.032	0.035	0.032
1	5	2	3.432	5.447	0.4956	-0.2978	3.430	5.445	0.496	-0.298	0.058	0.037	0.081	0.067
0.5	2	2	3.003	4.321	0.4477	-0.2580	3.005	4.320	0.448	-0.258	0.067	0.023	0.067	0.000

or

$$\Theta = \ln \frac{(\mu_0 - \mu)K - \varepsilon}{(\mu_0 - \mu)K + A_u}. \tag{3.11}$$

It then follows from (3.5) or (3.6) that

$$T = T_{u1} \left(\ln \frac{2\mu K e^{\Theta} + \mu_0 K - \mu K + \varepsilon}{\mu K + \mu_0 K - \varepsilon} \right)^{-1}, \tag{3.12}$$

or

$$T = T_{u1} \left(\ln \frac{2\mu K e^{\Theta} - \mu_0 K - \mu K + \varepsilon}{\mu K - \mu_0 K - \varepsilon} \right)^{-1}. \tag{3.13}$$

The dead time is thus

$$L = T\Theta. \tag{3.14}$$

The above development can be summarized as the following identification procedure.

Identification Procedure I   The biased relay experiment is performed. The process input responses $u(t)$ and output $y(t)$ are recorded, and the periods and the amplitudes of the oscillations are measured.

Step 1: Compute K from (3.9).

Step 2: Compute Θ from (3.10) or (3.11).

Step 3: Compute T from (3.12) or (3.13).

Step 4: Compute L from (3.14).

Table 3.2. Parameter Estimation from Biased Relay

	Process			Biased Relay				New Method			ATV Method		
Case	K	T	L	T_{u1}	T_{u2}	A_u	A_d	K	T	L	K	T	L
1	1	2	2	2.79	3.91	0.859	-0.480	1.000	1.999	2.002	1	1.658	2
2	1	1	3	3.50	4.18	1.241	-0.670	1.000	0.999	3.006	1	1.042	3
3	1	5	2	3.44	5.46	0.497	-0.299	0.999	4.990	2.009	1	4.068	2
4	1	5	1	2.15	3.65	0.318	-0.209	1.001	5.003	1.004	1	4.055	1

Simulation is carried out for processes with different normalized dead time to illustrate the accuracy of the proposed method. The amplitudes of biased relay are set at 1.3 and -0.7 respectively, and the hysteresis of relay is 0.1. The resultant limit cycles and model parameters are presented in Table 3.2. For comparison, the parameters obtained by the ATV (Luyben, 1987) are also given in Table 3.2, where it is assumed that the steady state gain is known and the dead time is read exactly. The Nyquist curves of the models and the corresponding real processes are shown in Figure 3.6. The results show that the proposed method can give nearly the exact identification of the process parameters.

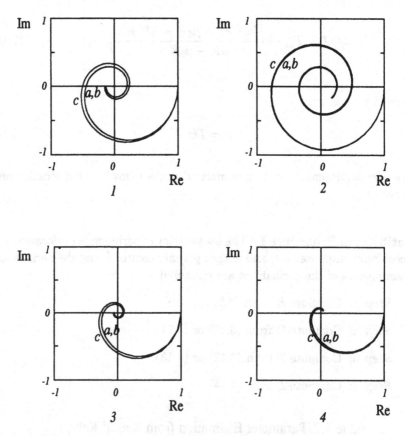

FIGURE 3.6. The Nyquist Curves of the Processes and the Models (a: real process; b: new method; c: ATV method)

In practice, many high-order processes may be approximated by first-order plus dead time models. The proposed method can also be used to model the high-order processes. The results for some typical processes are listed in Table 3.3. The Nyquist curves of the real processes and the models are shown in Figure 3.7, and they are very close to each other over phase range of 0 to $-\pi$.

Therefore, this low-order modelling will be accurate enough for control design in most cases.

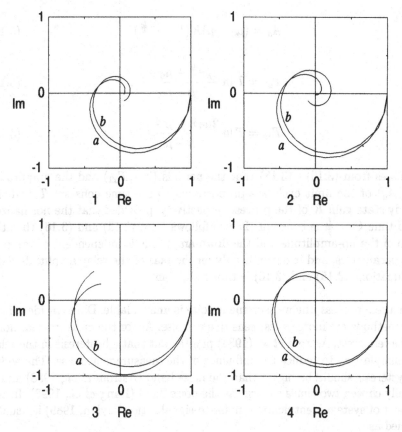

FIGURE 3.7. The Nyquist Curves of High-order Processes and Corresponding Low-order Models(a: real process; b: model)

If the relay has no hysteresis, i.e. $\varepsilon = 0$, (3.3)-(3.6) reduces to

Table 3.3. Models for the High-order Processes.

Case	Process	Model
1	$\frac{1}{(2s+1)^2}e^{-2s}$	$\frac{1.00}{4.072s+1}e^{-2.93s}$
2	$\frac{1}{(2s+1)^5}e^{-2s}$	$\frac{1.00}{6.809s+1}e^{-s}$
3	$\frac{1}{(s+1)(s^2+s+1)}e^{-0.5s}$	$\frac{1.00}{1.152s+1}e^{-2.1s}$
4	$\frac{-s+1}{(s+1)^5}e^{-s}$	$\frac{1.00}{2.99s+1}e^{-4.24s}$

$$A_u = (\mu_0 + \mu)K(1 - e^{-\frac{k}{T}}), \tag{3.15}$$

$$A_d = (\mu_0 - \mu)K(1 - e^{-\frac{k}{T}}), \tag{3.16}$$

$$T_{u1} = T \ln \frac{2\mu e^{\frac{k}{T}} + \mu_0 - \mu}{\mu + \mu_0}, \tag{3.17}$$

$$T_{u1} = T \ln \frac{2\mu e^{\frac{k}{T}} - \mu - \mu_0}{\mu - \mu_0}. \tag{3.18}$$

One sees from (3.15) - (3.18) that the periods (T_{u1}, T_{u2}) and the amplitudes (A_u, A_d) of the limit cycle are proportional to the time constant T and the steady state gain K of the process respectively, provided that the normalized dead time $\Theta = \frac{L}{T}$ is constant. It also follows from (3.15) and (3.16) that the ratio of the up-amplitude and the down-amplitude is independent of the process parameters, and it depends only on the bias of the relay amplitude. One of Equations (3.15) and (3.16) is thus redundant.

In a real process, the measurement noise is unavoidable. Different identification methods are more or less sensitive to noise. As for the measurement noise in the relay test, Astrom et al. (1984) pointed out that a hysteresis in the relay is a simple way to reduce the influence of the measurement noise. The width of hysteresis should be bigger than the noise band (Astrom et al., 1988) and is usually chosen two times as large as the noise band (Hang et al., 1993). In the context of system identification, noise-to-signal ratio (Haykin, 1989) is usually defined as

$$N_1 = \frac{mean\ power\ spectrum\ density\ of\ noise}{mean\ power\ spectrum\ density\ of\ signal},$$

which is Noise-to-Signal Power Spectrum Ratio, or

$$N_2 = \frac{mean(abs(noise))}{mean(abs(signal))},$$

which is Noise-Signal Mean Ratio.

The method was tested on the *Dual Process Simulator KI 100* described in Section 1.4.1. The following typical process is configured :

$$G(s) = \frac{1}{(2s + 1)^2} e^{-2s}.$$

Figure 3.8 shows the input-output responses for the biased relay tests with noise-to- signal ratio N_2 of 0.08 and 0.190 (N_1 of 0.006 and 0.033) respectively. The models are identified as

$$\hat{G}(s) = \frac{1.01}{3.53s+1}e^{-3.03s}, \qquad for\ N_2 = 0.08;$$

and

$$\hat{G}(s) = \frac{1.07}{3.28s+1}e^{-2.99s}, \qquad for\ N_2 = 0.190.$$

In this case, eight cycles of periodical stationary oscillations have been used in the steady- state gain calculation (3.9) and the up and down amplitudes are calculated as the mean value of those in the same data respectively. The Nyquist curves of the real processes and the models are shown in Figure 3.9, and they are very close to each other over phase range of 0 to $-\pi$. Since only stationary information is used to develop the model in our method and more periodical stationary oscillation data can be employed to effectively reduce noise effect on estimation, our method is less sensitive to noise compared with most identification methods.

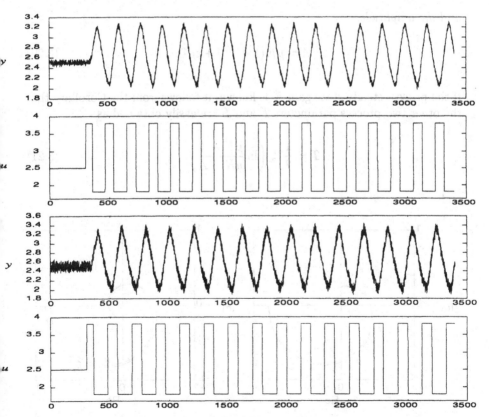

FIGURE 3.8. Input-output Responses of Biased Relay Experiments with Noise-to-signal Ratio N_2 of 0.08 and 0.190 respectively

3.3.3 A Special Case

The biased relay will cause the operating point to drift. It is thus a larger disturbance than an unbiased relay with the same amplitude. For this reason, a biased relay might sometimes be undesirable in practice. In this case, an unbiased relay has to be used and the process steady state gain is no longer identifiable. We now have to employ the first harmonic information contained in limit cycles instead of DC component. This makes exact identification from a single unbiased relay test still possible but at the expense of a bit more calculations. The following corollary, which can be easily reached with $\mu_0 = 0$ in the Theorem 3.1, gives the properties of the limit cycle oscillation under an unbiased relay feedback.

Corollary 3.1   *For the process of (3.2) under the relay feedback of Figure 3.2, the limit cycle oscillation is symmetric and is characterized by the period:*

$$T_\pi = 2T \ln \frac{2\mu K e^{\frac{L}{T}} - \mu K + \varepsilon}{\mu K - \varepsilon}, \tag{3.19}$$

and the amplitude:

$$A = \mu K (1 - e^{\frac{L}{T}}) + \varepsilon e^{\frac{L}{T}}. \tag{3.20}$$

If the process steady state gain K is known, it follows from the Corollary that the other two process parameters T and L are easily obtained as

$$T = \frac{1}{2} T_\pi \left(\ln \frac{\mu K + A}{\mu K - A} \right)^{-1}, \tag{3.21}$$

and

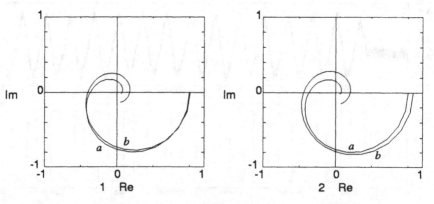

FIGURE 3.9. The Nyquist Curves of a Real Process and Corresponding Models (1: for $N_2 = 0.08$ 2: for $N_2 = 0.190$ a: real process b:model)

$$L = \frac{1}{2}T_\pi \left(\ln \frac{\mu K - \varepsilon}{\mu K - A} \right) \left(\ln \frac{\mu K + A}{\mu K - A} \right)^{-1}. \tag{3.22}$$

Otherwise, if K is unknown, we need an additional relation to determine the three parameters of the process in one relay test. Since the oscillation waveforms of the process input $u(t)$ and output $y(t)$ are periodic with the period T_π, they can be expanded into Fourier series. The first harmonics are then extracted and their coefficients give one point of process frequency response (Ramirez, 1985) via the formula:

$$G(j\omega_\pi) = \frac{\int_0^{T_\pi} y(t)e^{-j\omega_\pi t}dt}{\int_0^{T_\pi} u(t)e^{-j\omega_\pi t}dt} := A_0 e^{j\phi_0}, \tag{3.23}$$

where

$$\omega_\pi = \frac{2\pi}{T_\pi},$$

is the oscillation frequency. Unlike the describing function method, formula (3.23) introduces no approximation and gives an accurate estimation of one frequency response point at $\omega = \omega_\pi$. Substituting (3.2) into (3.23) and taking amplitudes on both sides give

$$\frac{K}{\sqrt{(\omega_\pi T)^2 + 1}} = A_0. \tag{3.24}$$

Combining (3.24) with (3.21) yields

$$e^{\frac{T_\pi}{2T}} = \frac{\mu A_0 \sqrt{(\omega_\pi T)^2 + 1} + A}{\mu A_0 \sqrt{(\omega_\pi T)^2 + 1} - A}. \tag{3.25}$$

This equation contains only one unknown T and it can be solved with a simple iterative method such as Newton interpolation. On the basis of (3.22) - (3.25), the following identification procedure is proposed.

Identification Procedure II The unbiased relay experiment is performed. The process input $u(t)$ and output $y(t)$ are recorded, and the period and the amplitude of the oscillations are measured.

    Step 1: Compute A_0 from (3.23).

    Step 2: Solve (3.25) with an iterative method to obtain T.

    Step 3: Compute K from (3.24).

    Step 4: Compute L from (3.22).

The same simulation examples as in Subsection 3.3.1 are used to illustrate the accuracy of the proposed method. The estimated model parameters are given in Table 3.4, and the Nyquist curves of the processes and the corresponding models are almost same and thus not shown. The results indicate that the proposed method can also give nearly the exact identification of the process parameters.

Table 3.4. Parameter Estimation from Unbiased Relay

Case	Process			Biased Relay			New Method			ATV Method		
	K	T	L	T_π	A	A_0	K	T	L	K	T	L
1	1	2	2	6.47	0.669	0.458	1.001	2.000	2.004	1	1.658	2
2	1	1	3	7.55	0.955	0.769	0.998	1.002	3.007	1	1.042	3
3	1	5	2	8.40	0.397	0.258	1.002	5.003	1.998	1	4.068	2
4	1	5	1	5.39	0.263	0.169	0.999	4.999	1.004	1	4.055	1

3.4   Relay Cascaded with a Nonlinear Mapping

In this section, a modified relay consisting of the basic relay cascaded to a nonlinear mapping function is presented. Through the appropriate use of non-linear mapping, certain desirable features not possible with the basic relay may be attained. These include better estimation accuracy, estimation of general and multiple points.

3.4.1   Improved Estimation Accuracy

While the relay feedback experiment design will yield sufficiently accurate results for many of the processes encountered in the process control industry, there are some potential problems associated with such techniques. These arise as a result of the approximations used in the development of the procedures for estimating the critical point, *i.e.* the ultimate frequency and ultimate gain. In particular, the basis of most existing relay-based procedures for critical point estimation is the describing function (DF) method. This method is approximate in nature, and under certain circumstances, the existing relay-based procedures could result in estimates of the critical point that are significantly different from their real values. Such problematic circumstances arise particularly in under-damped processes and processes with significant time-delay, and poorly tuned control loops would result if the critical point estimates were used for controller tuning.

Consider the relay feedback system of Figure 3.1. The usual method employed to analyze such systems is the describing function method which replaces the relay with an "equivalent" linear time-invariant system. For estimation of the critical point, the self-oscillation of the overall feedback system is of interest. Here, for the describing function analysis, a sinusoidal relay input,

$$e(t) = a \sin \omega t,$$

is considered, and the resulting signals in the overall system are analyzed. The relay output $u(t)$ in response to $e(t)$ would be a square wave having a frequency ω and an amplitude equal to the relay output level μ. Using a Fourier's series expansion, the periodic output $u(t)$ can be written as

$$u(t) = \frac{4\mu}{\pi} \sum_{k=1}^{\infty} \frac{\sin(2k-1)\omega t}{2k-1}.$$

The describing function of the relay $N(a)$ is simply the complex ratio of the fundamental component of $u(t)$ to the input sinusoid, $i.e.$

$$N(a) = \frac{4\mu}{\pi a}.$$

Since the describing function analysis ignores harmonics beyond the fundamental component, define here the residual ϱ as the entire sinusoidally-forced relay output minus the fundamental component, $i.e.$ the part of the output that is ignored in the describing function development,

$$\varrho = \frac{4\mu}{\pi} \sum_{k=2}^{\infty} \frac{\sin(2k-1)\omega t}{2k-1}.$$

In the describing function analysis of the relay feedback system, the relay is replaced with its quasi-linear equivalent DF, and a self-sustained oscillation of amplitude, a and frequency, ω_{osc} is assumed. Then, if $G(s)$ denotes the transfer function of the process, the variables in the loop must satisfy the following relations,

$$E = -Y,$$
$$U = N(a)E,$$
$$Y = G(j\omega_{osc})U.$$

This implies that it must follow

$$G(j\omega_{osc}) = -\frac{1}{N(a)}.$$

Relay feedback estimation of the critical point for process control is based on the key observation that the intersection of the Nyquist curve of $G(j\omega)$ and $-\frac{1}{N(a)}$ in the complex plane gives the critical point of the linear process. Hence, if there is a sustained oscillation in the system of Figure 3.1, then in the steady state, the oscillation must be at the ultimate frequency, i.e.

$$\omega_\pi = \omega_{osc},$$

and the amplitude of the oscillation is related to the ultimate gain, k_π by

$$k_\pi = \frac{4\mu}{\pi a}.$$

From the above discussion, it is evident that the accuracy of the relay feedback estimation depends on the residual ϱ which determines whether, and to what degree, the estimation of the critical point will be successful. For the relay, ϱ consists of all the harmonics in the relay output. The amplitude of the third and fifth harmonics are about 30% and 20% that of the fundamental component, and they are not negligible if fairly accurate analysis results are desirable, and therefore they limit the class of processes for which describing function analysis is adequate, i.e. the process must attenuate these signals sufficiently. This is the fundamental assumption of the describing function method which is also known as the *filtering hypothesis*. Mathematically, the hypothesis requires that the process, $G(s)$ must satisfy

$$|G(jk\omega_\pi)| \ll |G(j\omega_\pi)| , \quad k = 3, 5, 7, \cdots, \tag{3.26}$$

and

$$|G(jk\omega_\pi)| \to 0 , \quad k \to \infty. \tag{3.27}$$

Note that (3.26) and (3.27) require the process to be not simply low-pass, but rather low-pass at the ultimate frequency. This is essential as the delay-free portion of the process may be low-pass but the delay may still introduce higher harmonics within the bandwidth. Typical processes that fail the filtering hypothesis are processes with long time-delay and processes with resonant peaks in their frequency responses such that the undesirable frequencies are boosted instead of being attenuated. This provides one explanation for the poor results associated with these processes.

Having observed the accuracy problems associated with conventional relay feedback estimation, the design of a modified relay feedback that addresses the issue of improved estimation accuracy is considered next. Thus consider the modified relay feedback system of Figure 3.10 where the process is assumed to have the transfer function $G(s)$.

Define the mapping function f such that

$$u(t) = f(v(t)) = v_1(t) = \frac{4\mu}{\pi} \sin \omega t, \tag{3.28}$$

where $v_1(t)$ is the fundamental harmonic of $v(t)$, and μ is the amplitude of the relay element. For this modified relay feedback, it turns out that the following property may be stated:

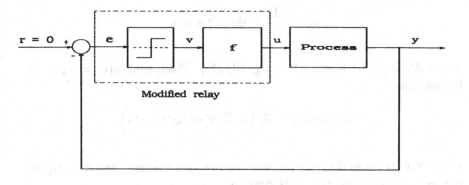

FIGURE 3.10. Modified relay feedback system.

Proposition 3.1   *Consider the use of the modified relay feedback defined by the system of Figure 3.10 and (3.28), where the process is assumed to have the transfer function $G(s)$. The set of signals*

$$u(t) = \frac{4\mu}{\pi} \sin(\omega^* t), \tag{3.29}$$

$$y(t) = \frac{4\mu}{\pi} A^* \sin(\omega^* t + \phi^*), \tag{3.30}$$

$$v(t) = N_r\left(-\frac{4\mu}{\pi} A^* \sin(\omega^* t + \phi^*)\right), \tag{3.31}$$

where $N_r(\cdot)$ denotes the relay function, describes an invariant set of the dynamical system defined by Figure 3.10 for

$$\phi^* = \arg\{G(j\omega^*)\} = -\pi,$$
$$A^* = |G(j\omega^*)|.$$

PROOF   Assume that the set-up of Figure 3.10 admits a solution of the form

$$u(t) = \frac{4\mu}{\pi} \sin(\omega^* t), \qquad (3.32)$$

for some ω^*. It will be shown that this solution is consistent with the definitions of the other signals in the loop so that this solution describes an invariant set of the dynamical system defined by Figure 3.10. With (3.32), it follows

$$y(t) = \frac{4\mu}{\pi} A^* \sin(\omega^* t + \phi^*),$$

where $A^* = |G(j\omega^*)|$ and $\phi^* = \arg\{G(j\omega^*)\}$. Then, since $e(t) = -y(t)$, it follows that

$$v(t) = N_r(e(t)) = N_r\left(-\frac{4\mu}{\pi} A^* \sin(\omega^* t + \phi^*)\right),$$

where $N_r(\cdot)$ denotes the relay function. Since this is the case, clearly $v_1(t)$, the fundamental harmonic of $v(t)$, is given by

$$v_1(t) = -\frac{4\mu}{\pi} \sin(\omega^* t + \phi^*).$$

This in turn implies that

$$u(t) = f(v(t)) = -\frac{4\mu}{\pi} \sin(\omega^* t + \phi^*). \qquad (3.33)$$

Equation (3.33) is consistent with (3.32) for

$$\phi^* = -(2n+1)\pi,$$

where n is a non-negative integer, and this thus characterizes a class of admissible solutions.

Therefore, for $n = 0$, the set of signals

$$
\begin{aligned}
u(t) &= \frac{4\mu}{\pi} \sin(\omega^* t), \\
y(t) &= \frac{4\mu}{\pi} A^* \sin(\omega^* t + \phi^*), \\
v(t) &= N_r\left(-\frac{4\mu}{\pi} A^* \sin(\omega^* t + \phi^*)\right),
\end{aligned}
$$

describes an invariant set of the dynamical system defined by Figure 3.10 with

$$\phi^* = -\pi,$$

as claimed. Note that the set (3.29)–(3.31) is clearly periodic in t.

Remark 3.1   The invariant set (3.29)–(3.31) established in Proposition 3.1 provides a suitable basis for estimation of the critical point for process control with improved accuracy. This is due to the fact that the analysis used to prove Proposition 3.1 does not depend on approximations so that theoretically, it is possible to calculate the critical point exactly from observations of the invariant set (3.29)–(3.31).

To estimate the critical point using the arrangement of Figure 3.10 and the result of Proposition 3.1, assume that an oscillation is observed which corresponds to the (admissible) solution

$$\phi^* = \arg\{G(j\omega^*)\} = -\pi.$$

This implies that

$$\omega^* = \omega_\pi.$$

Thus, the ultimate frequency may be obtained directly from the measurement of the frequency of the oscillation observed. For the critical point, the remaining parameter to be estimated is the ultimate gain, and from the measurements, this may be obtained from the ratio

$$k_\pi = \frac{a_\pi}{a_y} = \frac{4\mu}{\pi a_y},$$

where a_π and a_y are the observed amplitudes of the oscillations in $u(t)$ and $y(t)$ respectively. As in conventional relay feedback, the relay magnitude μ may be used as a design parameter to appropriately size the magnitude of the oscillations to handle situations with different levels of noise.

Remark 3.2   The analysis to prove Proposition 3.1 is essentially a time-domain analysis, and no approximations were involved in establishing the existence of the invariant set (3.29)–(3.31) and the frequency ω^* that characterized it. This may be considered as an improvement over existing methods which are based on describing function analysis, and which consequently yield estimation procedures for the critical point that involve approximations. The existing methods have the advantage of simplicity, and under certain circumstances, the improved procedure proposed here only yields small gains in accuracy. However, there are other circumstances (under-damped processes and processes with significant time-delay) where the gains in accuracy are significant.

It is interesting, actually, to note that the improved accuracy is also evident from the same describing function analysis used for the existing methods. Thus, a describing function analysis applied to the system of Figure 3.10 shows that corresponding to the input $e(t) = a \sin \omega t$, the nonlinearity's output $u(t)$ would also be a sinusoid described by

$$u(t) = \frac{4\mu}{\pi} \sin \omega t. \tag{3.34}$$

The describing function of the nonlinear system can then be obtained as

$$N(a) = \frac{4\mu}{\pi a}, \tag{3.35}$$

and it is quite straightforward to check that the residual $\varrho = 0$ so that the filtering hypothesis of the describing function analysis is satisfied strictly. It then follows that analysis using the different tools of describing functions also indicates that improved accuracy will be obtained from the procedure proposed here.

Implementation Procedures

The system described by Figure 3.10 and (3.28) defines the basic elements required for the proposed technique. For the construction of implementation procedures, the key point to note is that the module realizing the function f should be designed to extract the fundamental harmonic, and apply it as the signal $u(t)$. There are thus various practical procedures and variations that may be used.

For a simple practical implementation, one can obtain a rough estimate of the ultimate frequency from two or three switches of the relay under normal relay feedback, and then turn on the function f using the rough estimate. The function f is then updated iteratively using the observed signals. The steps in such a procedure would be as follows:

(a). Put the process under relay feedback (*i.e.* set $f(v(t)) = v(t)$) to obtain a rough estimate of its ultimate frequency $\hat{\omega}_\pi$ from m oscillations, say with $m = 2$ for example.

(b). Denote τ_n as the time corresponding to the nth switch of the relay to $v = \mu$, and $\hat{\omega}_{\pi,n-1}$ as the ultimate frequency estimate just prior to $t = \tau_n$. For $n > m$, update $f(\cdot)$ to output

$$u(t) = f(v(t)) = \frac{4\mu}{\pi} \sin \hat{\omega}_{\pi,n-1}(t - \tau_n).$$

(c). Obtain a new estimate of the ultimate frequency $\hat{\omega}_{\pi,n}$ from the resultant oscillation.

(d). Repeat (b)–(c) until successive estimates of the ultimate frequency show a satisfactory convergence.

Example 3.1   The results of a simulation example utilizing this procedure for implementing f are shown in Figure 3.11. In the simulation, the process is given by

$$G(s) = \frac{1}{s+1} e^{-10s},$$

and the set-up of Figure 3.10 is invoked. The signals $y(t)$, $u(t)$ and $v(t)$ are shown respectively in the top, middle and bottom frames in Figure 3.11.

With the above procedure for implementing f, conventional relay feedback experiment is held from $t = 0$ to $t = 52$ (i.e. $m = 2$). After the third relay switch to $v = \mu$ at $t = 52$, the function $f(\cdot)$ is initialized with the ultimate frequency estimate of $\hat{\omega}_\pi = 0.31$ obtained from the conventional experiment. Two subsequent updates of $f(\cdot)$ yield a resultant oscillation with a frequency which has essentially converged to the effectively exact ultimate frequency of $\omega_\pi = 0.29$. In the example, it can be seen that the simple method described above is relatively efficient and effective; it is also a complete and practically useful procedure for implementing the function $f(\cdot)$.

3.4.2 Estimation of a General Point

With the basic relay feedback approach, only one point on the process Nyquist curve is determined. It is possible, for example, to cascade a known linear dynamical system to the system in Figure 3.1 to obtain a frequency other than the ultimate frequency. For example, an integrator can be cascaded to obtain the point where the process Nyquist curve crosses the negative imaginary axis. Similarly, a first-order lag can be designed and cascaded to obtain a point with a frequency below the ultimate frequency. However, with these modifications, the frequency of interest cannot be specified; it is fixed by the choice of the linear element cascaded. Besides, the introduction of the linear system affects the amplitude response of the original process, and in the case of a gain reduction, a smaller signal-to-noise ratio (SNR) could affect the estimation accuracy adversely.

A variation of the basic relay will be presented in this subsection with a negative inverse describing function that is a ray through the origin in the third or fourth quadrant of the complex plane. With this particular nonlinearity, it is possible to obtain a point on the process frequency response at an arbitrarily specified phase lag. The system is flexible enough to give the frequency at a specified process lag of interest, $-\pi + \phi_m$ ($\phi_m \in [0, \pi)$), without affecting the amplitude response of the original process. The modified arrangement is similar to Figure 3.10 but with f defined such that

$$u(t) = f(v(t)) = v(t - L_a(\omega)).$$

where $L_a(\omega) = \frac{\phi_m}{\omega}$, is an adaptive time delay function, and ω is the oscillating frequency of $v(t)$.

The feature will be illustrated from a describing function analysis. Consider the modified relay in the dashed-box of Figure 3.10 which consists of a relay cascaded to the delay function, f. Corresponding to the reference input $e(t) = a\sin\omega t$, the output of the modified relay can be expanded using the Fourier series, and shown to be

$$u(t) = \frac{4\mu}{\pi} \sum_{k=1}^{\infty} \frac{\sin((2k-1)\omega t - \phi_m)}{2k-1}. \tag{3.36}$$

Hence, the describing function of the modified relay can be computed as

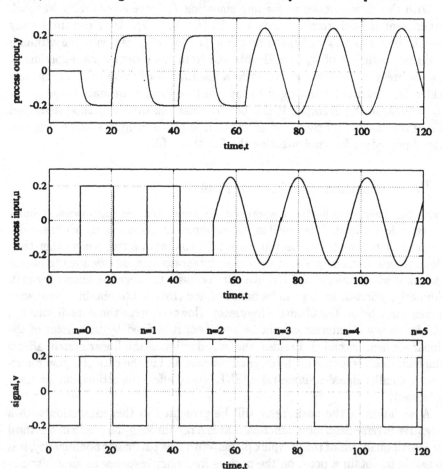

FIGURE 3.11. Example for illustration of the implementation procedure.

$$N(a) = \frac{4\mu}{\pi a}e^{-j\phi_m}. \tag{3.37}$$

The negative inverse describing function, $-\frac{1}{N(a)} = \frac{\pi a}{4\mu}e^{j(-\pi+\phi_m)}$ is thus a straight line segment through the origin as shown in Figure 3.12. By the feedback arrangement of Figure 3.10, the resultant amplitude and frequency of oscillation thus correspond to the intersection between $-\frac{1}{N(a)}$ and the process Nyquist curve. Hence, at the specified phase lag of $-\pi + \phi_m$, the inverse gain (k_ϕ) and the frequency of the process (ω_ϕ) can be obtained from the output amplitude (a) and frequency (ω_{osc}), i.e.

$$k_\phi = \frac{4\mu}{\pi a}, \tag{3.38}$$

$$\omega_\phi = \omega_{osc}. \tag{3.39}$$

With the arrangement of Figure 3.10, it is thus possible to automatically

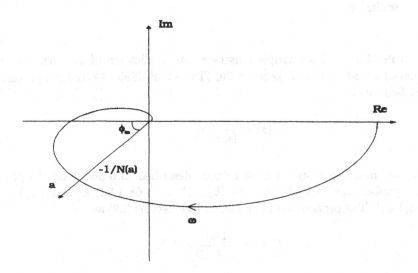

FIGURE 3.12. Negative inverse describing function of the modified relay.

track a frequency at the specified phase lag $-\pi + \phi_m$ for $\phi_m \in [0, \pi)$ without affecting the amplitude response of the original process. In this respect, it may be viewed as a generalized form of the relay feedback technique which tracks the frequency at the phase lag with $\phi_m = 0$. Besides, like relay feedback, the technique facilitates single-button tuning, a feature which is invaluable for autonomous/intelligent control applications.

Implementation Procedures

In the following, a practical implementation of procedure is presented to obtain two points on the process Nyquist curve ; $G(j\omega_\pi)$ at the $-\pi$ phase lag and $G(j\omega_\phi)$ at the specified phase lag of $-\pi + \phi_m$.

(a). Estimate the process ultimate gain $k_\pi = \frac{1}{|G(j\omega_\pi)|}$ and frequency ω_π with the normal relay feedback.

(b). With these estimates, an initial guess for the frequency $(\hat{\omega}_\phi)$ at the specified phase of $-\pi + \phi_m$ is

$$\hat{\omega}_\phi = \frac{\pi - \phi_m}{\pi}\omega_\pi,$$

and L_a is initialized as $L_a = \frac{\phi_m}{\hat{\omega}_\phi}$.

(c). Continue the relay experiment, and adapt the delay function f to the oscillating frequency ω_{osc}, i.e. $L_a = \frac{\phi_m}{\omega_{osc}}$. Upon convergence, $\omega_\phi = \omega_{osc}$, and $k_\phi = \frac{1}{|G(j\omega_\phi)|} = \frac{4\mu}{\pi a}$, where a is the amplitude of process output oscillation.

Example 3.2   This example illustrates the application of this method to enhanced auto-tuning of PID controller (Tan *et. al.* 1996). Consider a process described by

$$G(s) = \frac{1}{(s + 1)^2}e^{-0.5s}.$$

Using the modified relay feedback method described, two points on the process Nyquist curve are obtained as $(k_\pi, \omega_\pi) = (4.68, 1.92)$ and $(k_\phi, \omega_\phi) = (2.03, 1.02)$. The process can be modelled (see Section 3.6) as

$$\hat{G}(s) = \frac{1.01}{(s + 1)^2}e^{-0.5s}.$$

Based on this model, a PID controller can be designed and completed. The auto-tuning and closed-loop performance is shown in Figure 3.13. The improvement over a PID tuning method based on only the critical point is evident.

3.4.3   Estimation of Multiple Points

While the estimation procedure can give a general frequency of interest, tuning time is increased proportionally when more frequency estimations are required, especially if high accuracy is desirable. This is particularly true in the case of a process with a long time-delay where tuning time is considerably long. A further extension of procedure is possible which allows multiple frequency estimations in one single relay experiment. The arrangement and implementation is similar to Figure 3.10 but with the mapping function f defined such that

$$u(t) = f(v(t)) = \sum_{k=1}^{k_m} a_k \sin k\omega t \ , \quad k_m \in Z^+,$$

where $v_1 = a_1 \sin \omega t$ is the fundamental frequency of the input v, $a_k = \frac{4\mu}{k\pi}$ and $k_m\omega$ is the upper bound of the frequencies injected. The multiple points on the frequency response can then be estimated as

$$G(jk\omega_\pi) = \frac{\int_{-\frac{T_\pi}{2}}^{\frac{T_\pi}{2}} y(t)e^{-jk\omega_\pi t}dt}{\int_{-\frac{T_\pi}{2}}^{\frac{T_\pi}{2}} u(t)e^{-jk\omega_\pi t}dt} \ , \quad k = 1 \cdots k_m, \tag{3.40}$$

where $T_\pi = \frac{2\pi}{\omega_\pi}$.

Implementation Procedures

The steps in a practical implementation of the procedure would be as follows. For good estimation accuracy from this procedure, the estimation procedure is restricted to two frequencies, *i.e.* $k_m = 2$.

(a). Put the process under relay feedback (*i.e.* set $f(v(t)) = v(t)$) to obtain a rough estimate of its ultimate frequency $\hat{\omega}_\pi$ from m oscillations, say with $m = 2$ for example.

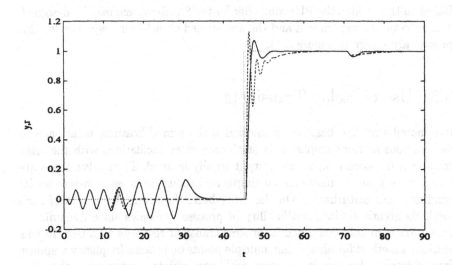

FIGURE 3.13. Auto-tuning session and closed-loop PID performance based on (– 2 points, - - - 1 point) – Example 3.2.

(b). Denote τ_n as the time corresponding to the nth switch of the relay to $v = \mu$, and $\hat{\omega}_{\pi,n-1}$ as the ultimate frequency estimate just prior to $t = \tau_n$. For $n > m$, update $f(\cdot)$ to output

$$u(t) = f(v(t)) = \frac{4\mu}{\pi} \sin \hat{\omega}_{\pi,n-1}(t - \tau_n) + \frac{2\mu}{\pi} \sin 2\hat{\omega}_{\pi,n-1}(t - \tau_n).$$

(c). Obtain a new estimate of the ultimate frequency $\hat{\omega}_{\pi,n}$ from the resultant oscillation.

(d). Repeat (b)–(c) until successive estimates of the ultimate frequency show a satisfactory convergence. The amplitude and phase of the resultant oscillation can be obtained from (3.40).

Example 3.3   This example illustrates the application of this method to auto-tuning of the Smith-Predictor Controller (Lee *et. al.* 1995b). Consider a high-order process given by

$$G(s) = \frac{1}{(s+1)^5} e^{-10s}.$$

Application of the modified relay method and the transfer function fitting of Section 3.6.2, the process model is obtained as

$$\hat{G}(s) = \frac{0.28}{s^2 + 0.98s + 0.28} e^{-11.57s}.$$

Based on this model, the PID controller for the Smith system may be designed. The auto-tuning experiment and the auto-tuned closed-loop responses for this process are shown in Figure 3.14.

3.5   Use of Relay Transients

It is noted that the basic relay method and its modifications mentioned in the previous sections employ only stationary relay oscillations with the relay transients in process input and output totally ignored. Their advantages are that the resultant estimations are simple and robust to noise, non-zero initial condition and disturbance. On the other hand, however, the nature of such methods greatly limits identifiability of process dynamics since dynamic information is mainly contained in process transient response to a test. In this section, a method for identifying multiple points on process frequency response from a single relay test is proposed with appropriate processing of the relay transient. Since the input and output transients resulting from a relay test cannot be directly transferred to frequency response meaningfully using FFT, a decay exponential is introduced to the process input and output so that the modified input and output will approximately decay to zero in a finite time interval. FFT is then employed to obtain the process frequency response. This method uses the same relay test as in the basic method and yet it can yield as many frequency response points as desired, and the estimation result is accurate.

3.5.1   The Method

Consider again the relay feedback system shown in Figure 3.1. If the process has a phase lag of at least π radians, the relay feedback will cause the system to oscillate. Assume that the system is at a steady-state before a relay feedback is applied at $t = 0$. The steady-state condition is an ordinary assumption which can also be found in other identification schemes such as a step test. For example, the Honeywell TDC 3000$^\times$ Robust PID Controller design requires a steady state at the beginning of a step identification test (Honeywell, 1995). In a relay test, the process output $y(t)$ and input $u(t)$ are recorded from the time the test starts until the system reaches a stationary oscillation.

In order to motivate our development, let us see why a direct application of the Fast Fourier Transform (FFT) to relay transient fails to get process frequency response. If the plant is described by $Y(s) = G(s)U(s)$, we formally have $G(j\omega) = Y(j\omega)/U(j\omega)$. The question is how to obtain $Y(j\omega)$ and $U((j\omega)$ from $y(t)$ and $u(t)$ numerically. At the first glance, it seems that $Y(j\omega)$ and $U(j\omega)$ can be easily obtained by taking FFT of $y(t)$ and $u(t)$ directly. Unfortunately, this operation is false and meaningless, as $y(t)$ and $u(t)$ are neither absolutely integrable nor periodic. To see the mechanism of this failure more clearly, we decompose $y(t)$ or $u(t)$ into the transient parts Δy or Δu and the

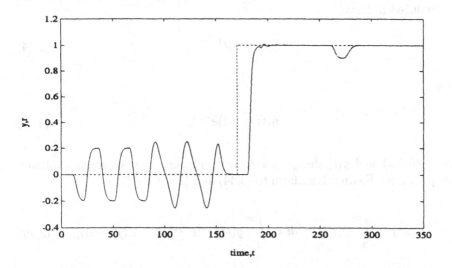

FIGURE 3.14. Auto-tuning and closed-loop performance – Example 3.3.

periodic stationary cycle parts y_s or u_s as

$$y(t) = \Delta y + y_s \tag{3.41}$$

and

$$u(t) = \Delta u + u_s \tag{3.42}$$

Once $y(t)$ and $u(t)$ reach the stationary oscillation status at $t = T_f$, both Δy and Δu are approximately zero afterwards. The Fourier transform of Δy is then

$$
\begin{aligned}
\Delta Y(j\omega) &= \int_0^\infty \Delta y(t) e^{-j\omega t} dt \\
&\approx \int_0^{T_f} \Delta y(t) e^{-j\omega t} dt.
\end{aligned}
\tag{3.43}
$$

Equation (3.43) can be computed at discrete frequencies with the standard FFT technique. FFT of one period of the periodic parts y_s (u_s) will actually give the scaled coefficients of the corresponding Fourier series of y_s (u_s) or the scaled amplitudes of the impulses of the extended Fourier transform of y_s or u_s (Cartwright, 1990). The FFT results from the transient parts and the periodic parts thus have different meanings and they cannot be added together. This means that the FFT cannot directly be applied to $y(t)$ or $u(t)$ to obtain the process frequency response.

To overcome the above obstacle, a decay exponential $e^{-\alpha t}$ ($\alpha > 0$) is here introduced to form

$$\tilde{y}(t) = y(t) e^{-\alpha t}, \tag{3.44}$$

and

$$\tilde{u}(t) = u(t) e^{-\alpha t}. \tag{3.45}$$

so that $\tilde{u}(t)$ and $\tilde{y}(t)$ decay to zero exponentially as t approaches infinity. Applying the Fourier Transform to (3.44) and (3.45) yields

$$\tilde{Y}(j\omega) = \int_0^\infty \tilde{y}(t) e^{-j\omega t} dt = \int_0^\infty y(t) e^{-\alpha t} e^{-j\omega t} dt = Y(j\omega + \alpha), \tag{3.46}$$

and

$$\tilde{U}(j\omega) = \int_0^\infty \tilde{u}(t)e^{-j\omega t}dt = \int_0^\infty u(t)e^{-\alpha t}e^{-j\omega t}dt = U(j\omega + \alpha), \qquad (3.47)$$

The integral intervals in equation (3.46) and (3.47) are infinite and digital computation of the infinite interval integration is not trivial. However, due to the introduction of the decay exponential $e^{-\alpha t}$, $\tilde{y}(t)$ and $\tilde{u}(t)$ are approximately zero after a certain time span. The infinite interval integration problem actually becomes a finite integration one. Thus, $\tilde{Y}(j\omega)$ can be computed at discrete frequencies with the standard FFT technique (Ramirez 1985). Suppose that $y(kT_s)$, $k = 0, 1, 2, \cdots, N - 1$, are samples of $y(t)$, where T_s is the sampling time interval. N is chosen such that $y((N - 1)T_s)$ has reached a stationary oscillation and the decay coefficient α is selected such that $\tilde{y}((N-1)T_s)$ formed from (3.44) has approximately decayed to zero. We then have

$$\begin{aligned}
\tilde{Y}(j\omega_i) &\approx T_s \sum_{k=0}^{\infty} \tilde{y}(kT_s)e^{-j\omega_i kT_s} \approx T_s \sum_{k=0}^{N-1} \tilde{y}(kT_s)e^{-j\omega_i kT_s} \\
&= FFT(\tilde{y}(kT_s)), \qquad i = 1, 2, \cdots, M,
\end{aligned} \qquad (3.48)$$

where $M = \frac{N}{2}$ and $\omega_i = 2\pi i/(NT_s)$. $\tilde{U}(j\omega)$ can be similarly computed by taking FFT of $\tilde{u}(kT_s)$. Therefore, the shifted process frequency response $G(j\omega_i + \alpha)$ is given by

$$G(j\omega + \alpha) = \frac{Y(j\omega_i + \alpha)}{U(j\omega_i + \alpha)} = \frac{\tilde{Y}(j\omega_i)}{\tilde{U}(j\omega_i)}, \qquad i = 1, 2, \cdots, M. \qquad (3.49)$$

Thus, the shifted process frequency response $G(j\omega + \alpha)$ is obtained. This identified shifted process frequency response is usually enough for process modelling and controller design (Wang et. al., 1997). However, if $G(j\omega)$ other than $G(j\omega + \alpha)$ is needed, we can first take the inverse FFT of $G(j\omega + \alpha)$ as

$$\tilde{g}(kT_s) := FFT^{-1}(G(j\omega_i + \alpha)) = g(kT_s)e^{-\alpha kT_s}. \qquad (3.50)$$

It then follows from $\tilde{g}(kT_s)$ that the process impulse response $g(kT_s)$ is

$$g(kT_s) = \tilde{g}(kT_s)e^{\alpha kT_s}. \qquad (3.51)$$

Applying the FFT again to $g(kT_s)$ would result in the process frequency response in $G(j\omega_i)$ form:

$$G(j\omega_i) = FFT(g(kT_s)). \qquad (3.52)$$

Unlike the method of Li *et. al.* (1991), one test is sufficient for the multiple-point frequency response identification and little prior knowledge of process is required with this method.

Parameter Selection   In order to implement this identification procedure in real time, we must give values for the decay factor α and the time span $T_f = (N - 1)T_s$ required in the FFT computations (3.48). Let us consider the selection of T_f first. Recall that our method can produce as many frequency response points as desired. Suppose that the number of the frequency response points to be identified from zero frequency to phase-crossover frequency ω_π is M. M should be specified by the system designers and will actually depend on the method which is used to tune the controller. It follows from (3.48) that the M frequency response points recovered by the FFT method are at the discrete frequencies $0, \Delta\omega, 2\Delta\omega, \cdots, (M - 1)\Delta\omega$, where $\Delta\omega = \omega_{i+1} - \omega_i = 2\pi/NT_s$, is the frequency increment. The definition of M means that $\omega_\pi \approx (M - 1)\Delta\omega$, thus

$$\omega_\pi \approx (M - 1)\frac{2\pi}{NT_s}. \tag{3.53}$$

On the other hand, one can measure the oscillation period T_π on line when a relay test is performed and ω_π can be estimated (Astrom, 1988) as

$$\omega_\pi \approx \frac{2\pi}{T_\pi}. \tag{3.54}$$

Equations (3.53) and (3.54) yield

$$N \approx (M - 1)\frac{T_\pi}{T_s} \tag{3.55}$$

where M should be specified a priori and should be large enough to ensure that the stationary oscillation is reached. The corresponding time span is then $T_f = (N - 1)T_s$.

For the decay coefficient α, its value should be such that $\tilde{y}(t)$ and $\tilde{u}(t)$ decay nearly to zero when the time approaches T_f, regardless of non-zero $y(t)$ and $u(t)$. It is this decay coefficient α that enables the infinite integral of Fourier transform to be approximately replaced by the finite digital integral in FFT. To have a good approximation in (3.48), the decay coefficient α should

$$e^{-\alpha T_f} \leq \varepsilon, \tag{3.56}$$

or

$$\alpha \geq -\frac{\ln \varepsilon}{T_f}, \tag{3.57}$$

where ε is the specified threshold and usually takes the value of $10^{-4} \sim 10^{-6}$.

Noise Issue   Noise is a big issue in system identification. It is apparent (Ljung, 1987) that in almost all identification methods a low noise-to-signal ratio is required. As for the measurement noise in the relay test, Astrom *et. al.* (1984) pointed out that a hysteresis in the relay is a simple way to reduce the influence of the measurement noise. The width of hysteresis should be bigger than the noise band (Astrom *et. al.*, 1988) and is usually chosen as two times larger than the noise band (Hang *et. al.*, 1993b). Filtering is another possibility (Astrom *et. al.*, 1984). The measurement noise is usually of high frequency while the process frequency response of interest for control analysis and design is usually in the low frequency region. We found through our experiments that a low pass filter can be employed to reduce the measurement noise effect further if the estimation of $\hat{G}(j\omega)$ is needed. But the estimation of $\hat{G}(j\omega + \alpha)$ is not sensitive to noise and no filter is needed if we stop with $\hat{G}(j\omega + \alpha)$. In particular, the process frequency response in $[0, \omega_\pi]$ is mostly critical for controller design. We found that in our experiments the measurement noise is indeed in the fairly high frequency region. Therefore, the cut-off frequency of the filter is determined with respect to the process frequency region of interest. The proposed method can reject noise quite effectively with the above anti-noise measures.

3.5.2 Simulation Study

The proposed frequency response identification method has been applied to several typical processes. For assessment of accuracy, the identification error is here measured by worst case error

$$ERR = \max_i \left\{ \left\| \frac{\hat{G}(j\omega_i) - G(j\omega_i)}{G(j\omega_i)} \right\| \times 100\%, i = 1, 2, \ldots, M \right\}, \qquad (3.58)$$

where $G(j\omega_i)$ and $\hat{G}(j\omega_i)$ are the actual and the estimated process frequency responses respectively. The Nyquist curve for phase ranging from 0 to $-\pi$ is being considered since this part is most significant for control design.

Example 3.4   This example is adopted from Li *et. al.* (1991) as

$$G(s) = \frac{e^{-2s}}{10s + 1}.$$

The model estimated by Li *et. al.* (1991) is $\hat{G}(s) = \frac{0.988e^{-2s}}{8.02s+1}$ and its identification error is $ERR = 22.32\%$. In Li's method, the dead-time is assumed to be known. For our method, a relay feedback is applied to the process. The process output and input are logged. $y(t)$ and $\tilde{y}(t)$ of the relay test are shown in Figure 3.15(a) while $u(t)$ and $\tilde{u}(t)$ are exhibited in Figure 3.15(b). The frequency response identified by the proposed method using MATLAB is shown in Figure 3.16. The ERR is 0.26%. This indicates that the proposed method provides a much more accurate process frequency response.

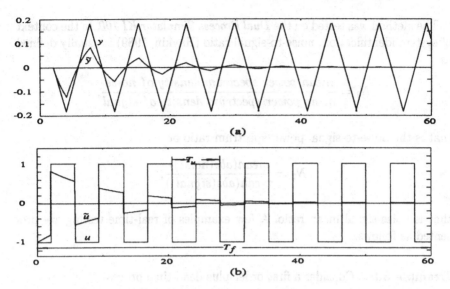

FIGURE 3.15. Signals under relay feedback

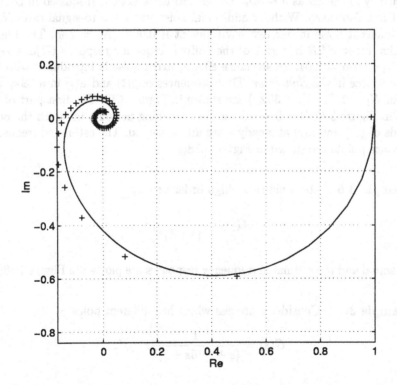

FIGURE 3.16. Nyquist plot of $G(j\omega)$. $\cdots$ actual, + estimated

The method was tested on the *Dual Process Simulator KI 100*. In the context of system identification, noise-to-signal ratio (Haykin, 1989) is usually defined as

$$N_1 = \frac{mean\ power\ spectrum\ density\ of\ noise}{mean\ power\ spectrum\ density\ of\ signal},$$

that is the noise-to-signal power spectrum ratio or

$$N_2 = \frac{mean(abs(noise))}{mean(abs(signal))},$$

that is noise-signal mean ratio. A few examples of real-time testing are presented as follows.

Example 3.5   Consider a first order plus dead-time process

$$G(s) = \frac{1}{5s+1}e^{-5s}.$$

The low pass filter is selected as a Butterworth low pass filter whose cut-off frequency is chosen as $3 \sim 5\omega_\pi$. Other anti-noise actions discussed in Section 3.5.1 are also taken. Without additional noise, the noise-to-signal ratio N_1 of the inherent noise in our test environment is 0.06% ($N_2 = 2\%$). The identification error *ERR* in terms of the shifted frequency response $G(j\omega + \alpha)$ is 2.77% with $\alpha = 0.04$. To see noise effects, extra noise is introduced with the noise source in the *Simulator*. Time sequences of $y(t)$ and $u(t)$ in a relay test under $N_1 = 10\%$ ($N_2 = 35\%$) are shown in Figure 3.17. The first part of the test in Figure 3.17 ($t = 0 \sim 30$) is the "listening period", in which the noise bands of $y(t)$ and $u(t)$ at steady state are measured. The estimated frequency response points are shown in Figure 3.18(a).

Example 3.6   For a multi-lag high order process:

$$G(s) = \frac{1}{(s+1)^8},$$

the actual and the estimated frequency responses are plotted in Figure 3.18(b).

Example 3.7   Consider a process which has different poles

$$G(s) = \frac{1}{(s+1)(5s+1)^2}e^{-2.5s}.$$

The actual and the estimated frequency responses are presented in Figure 3.18(c). The accuracy of the estimated frequency response is excellent.

Example 3.8   For a non-minimum phase plus dead time process

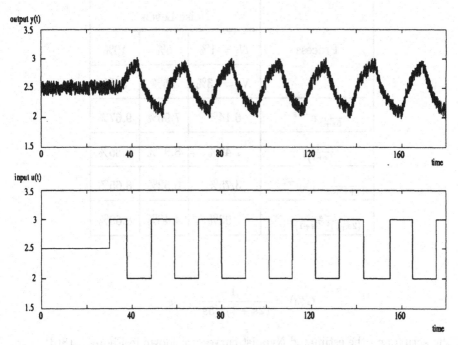

FIGURE 3.17. Real-time signals under relay feedback

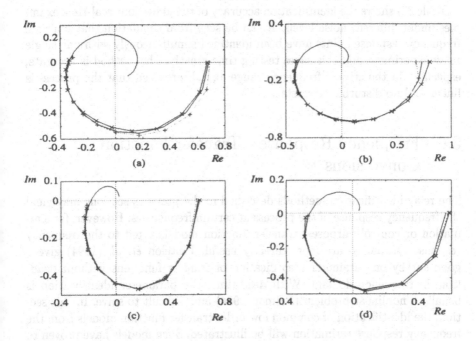

FIGURE 3.18. Nyquist Plot of $G(j\omega + \alpha)$.— actual, $\cdots$ + $\cdots$ estimated($N_1 = 1\%$), $\cdots$ × $\cdots$ estimated($N_1 = 10\%$).

Table 3.5. Identification Error (ERR)

Process	Noise Levels		
	$N_1 = 1\%$	6%	10%
	$N_2 = 13\%$	25%	35%
$\frac{1}{5s+1}e^{-5s}$	6.14%	7.96%	9.67%
$\frac{1}{(s+1)^8}$	7.42%	8.97%	9.85%
$\frac{1}{(s+1)(5s+1)^2}e^{-2.5s}$	3.75%	6.52%	8.65%
$\frac{1-s}{(2s+1)^4(5s+1)}e^{-s}$	2.93%	4.48%	8.69%

$$G(s) = \frac{1-s}{(2s+1)^4(5s+1)}e^{-s},$$

the actual and the estimated Nyquist curves are shown in Figure 3.18(d). The accuracy of the proposed method is evident.

Table 3.5 shows the identification accuracy of the above four real-time examples under different noise levels. It can be seen from simulation that multiple frequency response points have been identified simultaneously with one single relay experiment and this saves testing time greatly. The method is accurate, especially in the critical frequency range $[0, \omega_\pi]$, provided that the process is linear and no disturbance exists.

3.6   Frequency Response - Transfer Function Conversions

The relay identification methods described in the previous sections may yield the frequency response of the process at certain frequencies. However, for simulation or control purposes, transfer function models fitted to the frequency response estimation are more directly useful. Pintelon et. al. (1994) give a good survey on parameter identification of transfer functions without dead-time in frequency domain. With dead-time, the parameter identification is usually a nonlinear problem (Palmor, 1994) and difficult to solve. In this section, the identification of common low-order transfer function models from the frequency response estimation will be illustrated. Such models have proven to be relevant and adequate particularly in the process control industry. Algorithms are developed to calculate the parameters of these approximate transfer function models from the frequency response information acquired using the techniques discussed in the earlier sections. Such an identification technique is readily automated, and as such, it is useful for auto-tuning applications. Chapter 4 will use the method to obtain a first-order dead-time model for auto-tuning of FSA systems.

3.6.1 Single and Multiple Lag Processes

It is well known that many processes in the industry are of the low-order dynamics, and they can be adequately modeled by a rational transfer function of the following form :

$$\hat{G}(s) = \frac{K}{Ts+1}e^{-Ls}, \tag{3.59}$$

where K, T and L are real parameters to be estimated. It is desirable to distinguish this class of processes from higher-order ones since controller design and other supervisory tasks are often much simplified with the use of the model.

Two points on the process Nyquist curve are sufficient to determine the three parameters of (3.59). Assume that the following Nyquist points of the process $G(s)$ have been obtained: $G(j\omega_1)$ and $G(0)$, $\omega_1 \neq 0$. The process static gain is given by

$$K = G(0). \tag{3.60}$$

Equating the process and model at $\omega = \omega_1$ yields

$$|G(j\omega_1)| := \frac{1}{k_1} = \frac{K}{\sqrt{1+T^2\omega_1^2}}, \tag{3.61}$$

$$\arg G(j\omega_1) := \phi_1 = -\arctan T\omega_1 - L\omega_1. \tag{3.62}$$

It follows that

$$T = \frac{1}{\omega_1}\sqrt{k_1^2 K^2 - 1}, \tag{3.63}$$

$$L = -\frac{1}{\omega_1}(\phi_1 + \arctan\omega_1 T). \tag{3.64}$$

A more general model adequate for processes with a monotone open-loop step response is of the following form :

$$\hat{G}(s) = \frac{K}{(Ts+1)^n}e^{-Ls}, \tag{3.65}$$

where K, T and L are real parameters to be estimated, and $n \in Z^{+}$ is an order estimate of the process.

One method for order estimation based on this model was proposed by Lundh (1991). He noted that the maximal slope of the frequency response magnitude is a measure of the process complexity, and he uses the slope in the vicinity of the critical frequency to estimate the relative degree of the process, choosing one of three possible models to represent the process. In his method, FFTs are performed on the input and output of the process, from which the amplitudes of the first and third harmonics of the frequency spectrum are used to compute the amplitude gains at these two frequencies. From these frequencies, the slope of the frequency response magnitude at the geometrical mean of the harmonics is then calculated. A simpler method for order estimation is proposed here which does not require a Fourier transform on the input and output signals of the process.

Assume that the following Nyquist points of the process $G(s)$ have been obtained: $G(j\omega_1)$ and $G(j\omega_2)$, $\omega_1 \neq \omega_2$ and $\omega_1, \omega_2 \neq 0$. Equating the gain of the process and model at ω_1 and ω_2, one has

$$|G(j\omega_1)| := \frac{1}{k_1} = \frac{K}{\left(\sqrt{1+T^2\omega_1^2}\right)^n}, \qquad (3.66)$$

$$|G(j\omega_2)| := \frac{1}{k_2} = \frac{K}{\left(\sqrt{1+T^2\omega_2^2}\right)^n}. \qquad (3.67)$$

Simplifying (3.66) and (3.67), it follows that

$$\left[1 + \left(\frac{\omega_2}{\omega_1}\right)^2 \left((Kk_1)^{2/n} - 1\right)\right]^{n/2} = Kk_2. \qquad (3.68)$$

Equating the phase of the process and model at ω_1 and ω_2, it follows

$$\arg G(j\omega_1) := -\phi_1 = -n\arctan T\omega_1 - L\omega_1, \qquad (3.69)$$
$$\arg G(j\omega_2) := -\phi_2 = -n\arctan T\omega_2 - L\omega_2 \qquad (3.70)$$

A simple algorithm to obtain the parameters in (3.65) is outlined below. $\bar{n}$ is a specified upper bound on the order of the process.

- Iterate from $n = 1$ to $n = \bar{n}$.

    - Compute K from (3.68), T from (3.66) or (3.67), and L from (3.69).
    - Compute the cost function derived from (3.69),

$$J(n) = |-\phi_2 + n\arctan T\omega_2 + L\omega_2|$$

with the values of n, K, T and L above.

- At the end of the iteration, a suitable set of model parameters (n, K, T, L) corresponds to $n = n_{min}$, where

$$J(n_{min}) = \min_n \{J(n)\}.$$

3.6.2 Second-order Modelling

In this subsection, a non-iterative method is proposed for the following stable low-order plus dead-time model:

$$\hat{G}(s) = \frac{1}{as^2 + bs + c}e^{-Ls}, \tag{3.71}$$

which can represent both monotonic and oscillatory processes.

Transfer Function Modelling from $G(j\omega)$

Assume that the process frequency response $G(j\omega_i)$, $i = 1, 2, \cdots, M$, is available, and it is required to be fitted into $\hat{G}(s)$ in (3.71) such that

$$G(j\omega_i) = \frac{1}{(j\omega_i)^2 a + j\omega_i b + c}e^{-j\omega_i L}, \quad i = 1, 2, \cdots, M. \tag{3.72}$$

The determination of the parameters a, b, c and L in (3.72) seems to be a nonlinear problem. One way of solving this problem is to find the optimal a, b and c given L and iteratively determine L by some searching algorithm. To avoid the iteration, we take the magnitude of both sides of (3.72) as

$$\begin{bmatrix} \omega_i^4 & \omega_i^2 & 1 \end{bmatrix} \theta = \frac{1}{|G(j\omega_i)|^2}, \quad i = 1, 2, \cdots, M, \tag{3.73}$$

where $\theta = \begin{bmatrix} a^2 & b^2 - 2ac & c^2 \end{bmatrix}^T$. This action shields the effects from L and forms a system of linear equations (3.73) in θ which can be solved for θ with the linear least squares method. Then, $[a\ b\ c] = \begin{bmatrix} \sqrt{\theta_1} & \sqrt{\theta_2 + 2\sqrt{\theta_1\theta_3}} & \sqrt{\theta_3} \end{bmatrix}$. In addition, the phase relation in (3.72) gives

$$\omega_i L = -arg[G(j\omega_i)] - tan^{-1}\left(\frac{b\omega_i}{c - a\omega_i^2}\right), \quad i = 1, 2, \cdots, M. \tag{3.74}$$

Obviously, L can be obtained again with the least squares method.

Transfer Function Modelling from $G(j\omega + \alpha)$

With the method in Section 3.5, $G(j\omega + \alpha)$ is a more direct product than $G(j\omega)$. Thus, direct transfer function modelling from $G(j\omega + \alpha)$ is of more interest in this case. Determining the parameters a, b, c and L from $G(j\omega + \alpha)$ is not an easy job. Here, we present a possible solution.

When the process frequency responses $G(j\omega + \alpha)$, $i = 1, 2, \cdots, M$ are available, they are fitted into $\hat{G}(s)$ in (3.71) such that

$$G(j\omega_i + \alpha) = \frac{1}{(j\omega_i + \alpha)^2 a + (j\omega_i + \alpha)b + c} e^{-(j\omega_i + \alpha)L}, \quad i = 1, 2, \cdots, M.$$
(3.75)

Taking the magnitude of both sides of (3.75) yields

$$|G(j\omega_i + \alpha)| = \frac{1}{\sqrt{(\nu - a\omega_i^2)^2 + \omega_i^2(b + 2a\alpha)^2}} e^{-\alpha L}, \quad i = 1, 2, \cdots, M. \quad (3.76)$$

where $\nu = a\alpha^2 + b\alpha + c$. We first compute $\hat{a}, \hat{b}$ and $\hat{\nu}$ from

$$|G(j\omega_i + \alpha)| = \frac{1}{\sqrt{(\hat{\nu} - \hat{a}\omega_i^2)^2 + \omega_i^2(\hat{b} + 2\hat{a}\alpha)^2}}, \quad i = 1, 2, \cdots, M. \quad (3.77)$$

Equation (3.77) can be changed into is a system of linear equations in $\hat{\theta}$ as

$$\begin{bmatrix} \omega_i^4 & \omega_i^2 & 1 \end{bmatrix} \hat{\theta} = \frac{1}{|G(j\omega_i + \alpha)|^2}, \quad i = 1, 2, \cdots, M, \quad (3.78)$$

where $\hat{\theta} = \begin{bmatrix} \hat{a}^2 & \hat{b}^2 + 4\hat{a}\hat{b}\alpha + 4\hat{a}^2\alpha^2 - 2\hat{a}\hat{\nu} & \hat{\nu}^2 \end{bmatrix}^T$. Equation (3.78) can be solved with the linear least squares method and we have $\hat{a}, \hat{b}$ and $\hat{\nu}$. Equations (3.76) and (3.77) yield

$$\frac{1}{\sqrt{(\nu - a\omega_i^2)^2 + \omega_i^2(b + 2a\alpha)^2}} e^{-\alpha L} = \frac{1}{\sqrt{(\hat{\nu} - \hat{a}\omega_i^2)^2 + \omega_i^2(\hat{b} + 2\hat{a}\alpha)^2}} \quad (3.79)$$

for every ω_i, $i = 1, 2, \cdots, M$. Matching the coefficients of like power of ω_i in (3.79) yields

$$\begin{cases} a = e^{-L\alpha}\hat{a} \\ b = e^{-L\alpha}\hat{b} \\ c = e^{-L\alpha}\hat{c} \end{cases} \quad (3.80)$$

for stable process and $\alpha > 0$. Phase relation of (3.75) leads to

$$\omega_i L = -\arg[G(j\omega_i + \alpha)] - \tan^{-1}(\frac{b\omega_i + 2a\alpha\omega_i}{\nu - a\omega_i^2}), \quad i = 1, 2, \cdots, M, \quad (3.81)$$

Table 3.6. Transfer Function Modelling

Process	Process Models	
	from $G(j\omega)$	from $G(j\omega + \alpha)$
$\frac{1}{10s+1}e^{-2s}$	$\frac{1}{(9.99s+1)}e^{-2.01s}$	$\frac{1}{(10.00s+1)}e^{-2.00s}$
$\frac{1}{(s+1)^{10}}$	$\frac{1}{8.29s^2+5.06s+1}e^{-5.02s}$	$\frac{1}{8.16s^2+5.05s+1}e^{-5.02s}$
$\frac{-s+1}{(s+1)^5}e^{-2s}$	$\frac{1}{2.68s^2+3.04s+0.999}e^{-4.90s}$	$\frac{1}{2.63s^2+3.04s+1.01}e^{-4.91s}$
$\frac{1}{s^2+s+1}e^{-s}$	$\frac{1}{s^2+s+1}e^{-1.00s}$	$\frac{1}{1.0s^2+1.0s+1.0}e^{-1.00s}$

Using (3.80), Equation (3.81) becomes

$$\omega_i L = -\arg[G(j\omega_i + \alpha)] - tan^{-1}(\frac{\hat{b}\omega_i + 2\hat{a}\alpha\omega_i}{\hat{\nu} - \hat{a}\omega_i^2}), \quad i = 1, 2, \cdots, M, \qquad (3.82)$$

where $\hat{a}, \hat{b}$ and $\hat{\nu}$ have been obtained by (3.78). L then can be obtained with the least squares method. Once L is known, a, b and c in (3.75) can then be determined by (3.80).

Simulation is carried out on a number of processes to illustrate our transfer function modelling procedures. The resultant transfer functions derived from the estimated $G(j\omega)$ and $G(j\omega+\alpha)$ are listed in Table 3.6, which shows that the two methods produce very similar results. The actual and estimated Nyquist plots are shown in Figure 3.19(a)-Figure 3.19(d) and they fit each other very well.

3.6.3   First-order Unstable Process with Time-delay

Consider a first-order open-loop unstable process with a time-delay described by

$$\hat{G}(s) = \frac{K}{Ts-1}e^{-sL}. \qquad (3.83)$$

Assume the Nyquist point $G(j\omega_1)$ of the process is known where $\omega_1 \neq 0$, and an estimate of its time-delay L is available. Equating the process and the model (3.83) at the frequency ω_1, it follows that

$$|G(j\omega_1)| \quad := \quad \frac{1}{k_1} = \frac{K}{\sqrt{(T\omega_1)^2 + 1}}, \qquad (3.84)$$

$$\arg G(j\omega_1) \quad := \quad -\phi_1 = -\arctan(-T\omega_\pi) - \omega_\pi L \qquad (3.85)$$

Simplifying (3.84) and (3.85), the remaining two parameters of (3.83) are obtained as

$$T = \frac{1}{\omega_1}\tan(\omega_1 L - \phi_1), \qquad (3.86)$$

$$K = \frac{\sqrt{(T\omega_1)^2 + 1}}{k_1}. \qquad (3.87)$$

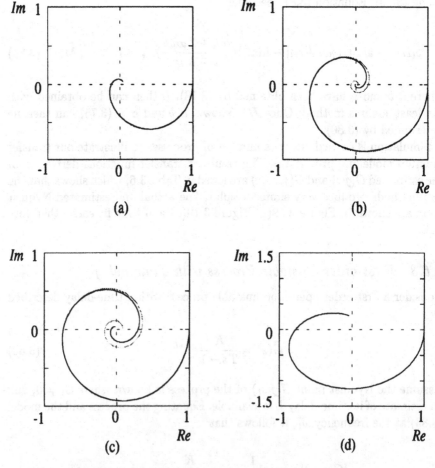

FIGURE 3.19. The Nyquist Plots: — actual, $\cdots$ estimated

3.7   Problems

1. Consider a process described by the following transfer function :

$$G(s) = \frac{Y(s)}{U(s)} = \frac{2}{10s + 1}e^{-20s}.$$

Write a simulation program to implement a relay feedback experiment and estimate the ultimate gain and frequency of the process. Implement a suitable modification to the classical setup to obtain more accurate estimates of these process characteristics.

2. Consider a process described by the following transfer function :

$$G(s) = \frac{Y(s)}{U(s)} = \frac{1}{(2s + 1)^5}e^{-3s}.$$

Write a simulation program to implement a suitable experiment to obtain the parameters of a first-order transfer function.

3. What advantages does a relay offer for process identification?

4. Use the algorithm in Subsection 3.6.2 to obtain the second-order approximation to the following transfer function:

$$G(s) = \frac{Y(s)}{U(s)} = \frac{1}{(s^2 + s + 1)(2s + 1)^5}e^{-3s}.$$

5. In Figure 3.4, the relay is biased in terms of the relay amplitude, and this enables static gain estimation. It is equally possible to bias a standard relay in terms of the relay width for such an estimation. Derive accordingly the estimation formulas for first-order modelling.

6. The robustness of frequency response estimation with respect to noise can be much improved by using more cycles of relay oscillations. Provide qualitative and quantitative explanations for such robustness improvement.

7. Can the process deadtime be estimated from relay oscillations using the output time response? How can this be done and what is the accuracy and robustness?

8. Relay feedback can provide sufficient excitation at the process critical frequency and thus yields very accurate and robust estimation of that point. But its excitation at frequencies far away from that point is much weaker and poor estimation at these frequencies can be expected. Essentially, we should increase the excitation at zero frequency and the frequency with phase lag of 90 degree to obtain uniform and good estimation of the process frequency response over the most important frequency range from zero frequency to critical frquency. Design a composite test (possibly with relay, step, and relay modifications) for such a purpose.

Controller Design and Tuning

This chapter presents Controller Design for the FSA system using a first-order model. This model is well-known to be adequate for a wide range of process dynamics encountered in industrial processes and controller realization using this model is simple and efficient. Exceptions are poorly damped and unstable processes for which other variants of transfer function may have to be considered. A tuning methodology using the relay feedback methods of Chapter 3 enables this Controller Design on an automated platform, without *a prior* information of the process dynamics. Subsequent and continuous self-tuning of the controller is realized using FFT techniques on opportune transients upset by set-point changes or load disturbances.

4.1   Controller Design

Consider a process characterized by a first-order plus dead time transfer function:

$$G_p(s) = \frac{K}{Ts+1}e^{-Ls}. \tag{4.1}$$

This model may be obtained by the relay-based identification method and frequency response - transfer function conversion techniques described in Chapter 3. The FSA will here be adopted to control the process.

Theoretically, the FSA will eliminate the time-delay from the closed-loop characteristic equation, and the control performance is dominated by the desirable closed-loop polynomial $p(s)$. Recall from section 2.3 of Chapter 2 that with the first-order modelling (4.1), the generalized process is defined as

$$G(s) = \frac{a_o}{s(s+b_1)}e^{-Ls}, \tag{4.2}$$

where $a_o = -\frac{K}{T}$ and $b_1 = \frac{1}{T}$. Since it is a second-order system, $p(s)$ is chosen as a degree 2 polynomial:

$$p(s) = s^2 + 2\zeta\omega_o s + \omega_o^2, \tag{4.3}$$

where ω_o and ζ are respectively the natural frequency and the damping factor of the desired closed-loop response. In order to simplify design, the observer polynomial $q(s)$ is chosen similar to the process denominator:

$$q(s) = s + \frac{1}{T}, \tag{4.4}$$

and $h(s)$ is chosen as $h(s) = h_1(s+b_1)$. Then the polynomial equation of (2.5):

$$k(s)b(s) + h(s)a(s) = q(s)f_L(s)$$

can be solved, and the solution is

$$k(s) = \left(\frac{1}{T} - 2\omega_o\zeta + \omega_o^2 T\right)e^{-\frac{s}{T}} - \omega_o^2 T, \qquad (4.5)$$

and

$$h(s) = \frac{T\omega_o^2}{K}s + \frac{\omega_o^2}{K}. \qquad (4.6)$$

All the required polynomials $p(s)$, $q(s)$, $k(s)$ and $h(s)$ to implement the FSA controller are now available from (4.3)-(4.6) respectively. The ideal closed-loop transfer function can be shown to be

$$G_{yr}^*(s) = \frac{\omega_o^2}{s^2 + 2\zeta\omega_o s + \omega_o^2}e^{-Ls}.$$

In this case, the closed-loop zero contributed by the polynomial $h(s)$ has been cancelled by the pole contributed by the observer polynomial $q(s)$, and does not appear in $G_{yr}^*(s)$.

Where uncertainty bounds of the model is known (or specified), ω_o and ζ are chosen to satisfy robustness conditions (see Chapter 5). Otherwise, as a rule-of-thumb, ω_o and ζ may be specified subject to $\frac{1}{\zeta\omega_o} = \lambda T$, where λ reflects the closed-loop response speed relatively to the process time constant, and it usually takes its value between 0.5 and 1.0. ζ is chosen according to the required closed-loop overshoot, and is usually in the range of 0.5 to 1.

4.2   Auto-tuning

The Controller Design procedure may be evoked once the model is available and desired closed-loop behaviour specified. An effective automatic tuning of the FSA system will therefore involve extracting the process dynamics information from simple and efficient experiments. Several approaches are possible. Traditional step analysis and frequency response analysis are known to produce simple transfer functions of the model through a curve fitting process. In this section, however, the variety of relay-based methods presented in Chapter 2 will be considered instead, due to the good properties associated. The end-product will be one of the common transfer function models of the process presented in Chapter 2. In this section, the first-order model will be considered, and several modelling approaches are feasible.

One approach would be to obtain the critical point of the process using the conventional relay method. The enhanced method of Section 3.4.1 may be used to provide better accuracy. With the critical point, Ziegler-Nichols like rules can be used to tune PI control for unity feedback systems. A subsequent set-point change enables the static gain of the process to be obtained as

$$K = \frac{\Delta y_{ss}}{\Delta u_{ss}}, \qquad (4.7)$$

where Δy_{ss} and Δu_{ss} are the changes in steady-state input and output signals corresponding to the set-point change. Another approach is to use the method in section 3.5 of Chapter 3 to obtain $g(0)$ and $g(j\omega_1)$ with $\omega_1 \neq 0$. Sufficient information is now available to apply the first algorithm of Section 3.6.1 to compute the model.

In the following, several simulation examples for different kinds of dynamics will be presented to illustrate the applicability and effectiveness of the auto-tuning method. The first approach described will be illustrated. This simulation is a little unusual as one sees from (2.10) that the controller contains a convolution integral over a finite time, which cannot be realized by a rational function.

Example 4.1   Consider a process with a single first-order lag and time-delay,

$$G_p(s) = \frac{1}{s+1}e^{-4s}.$$

After a relay feedback experiment, the ultimate gain and period are estimated as $(k_u, t_u) = (1.29, 9.93)$. Using Hagglund's tuning rules for delay systems (Hagglund, 1991), a PI controller is commissioned in a conventional unity feedback system with $(k_c, \tau_i) = (0.32, 2.52)$. With this PI controller commissioned, the next set point change leads to the natural estimate of $K = 1.0$. From the algorithm of Section 3.6.1, the model parameters can be computed as $T = 1.28$ and $L = 3.89$. Thus, the identified model is

$$\hat{G}_p(s) = \frac{1}{1.28s+1}e^{-3.89s}.$$

ζ and λ are specified as 1 and 0.5 respectively, and ω_o is then calculated as 1.56. $q(s)$, $k(s)$ and $h(s)$ are obtained from (4.4)-(4.6) as

$$q(s) = s + 0.78,$$
$$k(s) = -3.12,$$
$$h(s) = 3.12s + 2.4.$$

The FSA control system design is thus completed. This auto-tuning process is shown in Figure 4.1. The first part from $t = 0$ to $t = 30$ exhibits the relay control. The PI controller is commissioned at $t = 30$, and at $t = 38$, the system is already in the steady state. A set point change occurs at $t = 40$, and the FSA controller is commissioned at $t = 85$. Subsequently, the much improved set point response over the PI control from $t = 90$, and the load response from $t = 120$, are observed.

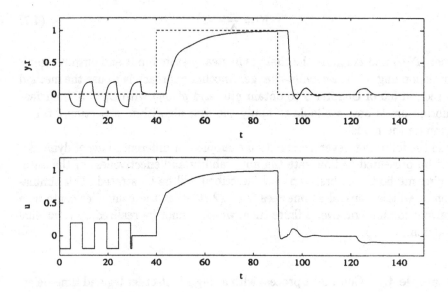

FIGURE 4.1. FSA auto-tuning and closed-loop performance – Example 4.1.

Example 4.2   Consider a high vacuum distillation column with transfer function between the viscosity and the reflux flow given by

$$G_p(s) = \frac{0.0077}{s^2 + 0.2326s + 0.0135}e^{-18.7s}.$$

After a relay feedback and closed-loop step test, the model is identified as

$$\hat{G}_p(s) = \frac{0.57}{1 + 12.72s}e^{-23.2s}.$$

The auto-tuning procedure and the subsequent closed-loop performance is shown in Figure 4.2.

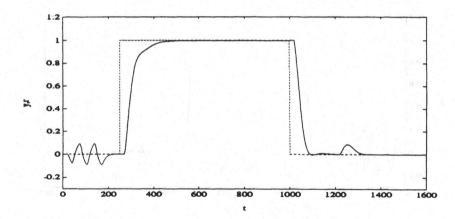

FIGURE 4.2. FSA auto-tuning and closed-loop performance – Example 4.2.

Example 4.3   Consider a high-order system:

$$G_p(s) = \frac{1}{(1+s)^n}, \quad n = 10, 20.$$

With auto-tuning, the models are identified respectively as

$$\hat{G}_p(s) = \frac{1}{1+2.72s}e^{-7.69s},$$

for $n = 10$, and

$$\hat{G}_p(s) = \frac{1}{1+4.95s}e^{-15.67s},$$

for $n = 20$. The auto-tuning and the subsequent performance is shown in Figure 4.3.

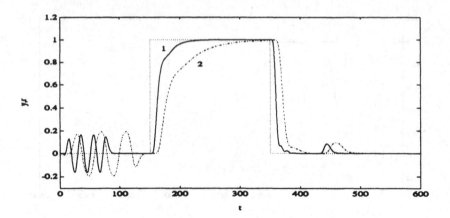

FIGURE 4.3. FSA auto-tuning and closed-loop performance (1– $n = 10$; 2 – $n = 20$) – Example 4.3.

Example 4.4   Consider the non-minimum phase system:

$$G_p(s) = \frac{1 - \beta s}{(1 + s)^3}, \quad \beta = 1, 1.5.$$

Using auto-tuning, the identified models are, respectively,

$$\hat{G}_p(s) = \frac{1}{1 + 1.61s} e^{-2.25s},$$

for $\beta = 1$, and

$$\hat{G}_p(s) = \frac{1}{1 + 1.01s} e^{-2.89s},$$

for $\beta = 1.5$. The auto-tuning and the subsequent performance is shown in Figure 4.4.

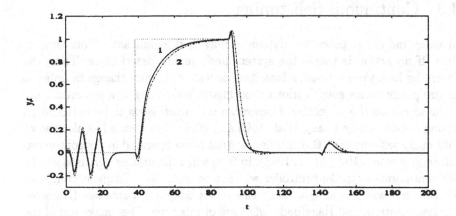

FIGURE 4.4. FSA auto-tuning and closed-loop performance (1– $\beta = 1$; 2 – $\beta = 1.5$)
– Example 4.4 .

These simulation examples show that significant performance improvement
over PI control is achievable through use of the auto-tuned FSA for processes
with large time delay.

Case study   The auto-tuning procedure described above is applied to real-
time experiments on level control in the coupled-tanks system with transport
delay described in Section 1.4.2, and some results will be briefly described here.

Figure 4.5 shows that a relay feedback experiment is carried out from $t = 0$
to $t = 83.3$ from which the process critical point is estimated as $(k_u, t_u) =$
$(1.02, 19.87)$. Using Hagglund rules, the PI parameters to be used in a conventional unity feedback system are $(K_p, T_I) = (0.25, 5.07)$. With the PI commissioned, the next set-point change at $t = 106$ results in an estimate of the
process static gain, $K = 1.09$. Thus, the model is identified using the algorithm
of Section 3.6.1 as

$$\hat{G}_p(s) = \frac{1.09}{1.55s + 1} e^{-8.5s}.$$

ζ and λ are chosen as 1 and 0.5 respectively. This leads to $\omega_o = 1.2$. $q(s)$, $k(s)$
and $h(s)$ are then computed as

$$q(s) = s + 0.64,$$
$$k(s) = -2.24,$$
$$h(s) = 2.04s + 1.32.$$

The FSA controller is commissioned at $t = 200$. A set-point step change is
introduced at $t = 200$, and a load disturbance occurs at $t = 233$. One sees from
Figure 4.5 that the FSA controller behaves much better than a conventional
PI controller.

4.3   Continuous Self-tuning

In many industrial processes, dynamics may drift significantly from time to time. If no action is taken, the system performance deteriorates. Thus, the controller being used should adapt itself to such dynamics change in order to ensure performance specifications. Continuous self-tuning is a powerful technique to realize this objective. However, a self-tuner needs to be switched off during a load change (Hang *et al.*, 1993 and 1995). Experience has also shown that many self-tuning PID controllers are not being operated in the continuous self-tuning mode (Hang *et al.*, 1993). In fact, when the process perturbs slowly, the continuous adaptive controller will not be necessary. Frequent controller adjustments will be enough. Some rule-based adaptation schemes (Hang *et. al.*, 1993; Astrom and Hagglund, 1995) are of this type. They make use of the process response under a set-point change or a load disturbance to re-tune their controllers. Based on the process response patterns, the controllers will be modified using the pre-setting rules. These rules are generally quite complex in order to work efficiently in different situations and are not easy to devise. The rules are usually empirical and the resultant performance mainly depends on the rule-formulator experience. These methods are more suitable for those controllers, which allow rather small successive changes in the controllers parameters.

To have a more accurate modelling and a better control performance, a new adaptive control scheme from nature responses such as a step-point response and a load disturbance response is here proposed. Auto-tuning is used only to start up the controller. The on-line adaptive controller will wait for a significant output response to occur. If the response is due to a set-point change, then the identification method in section 3.5 of Chapter 3 and the FSA design in Section 4.1 of Chapter 4 can be exploited to update the controller. This is straightforward and no further discussion will be given here. It should be pointed out that a set-point change is seldom made in process industry and control adaptation from a load disturbance response is much more demanding.

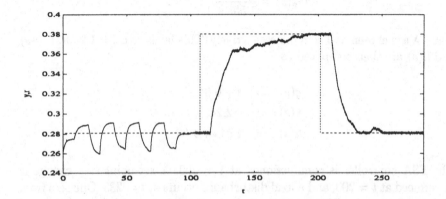

FIGURE 4.5. FSA auto-tuning and closed-loop performance – Case study.

If there is no set-point change, then any output response must be caused by some load disturbance. The problem is how to use the resultant transients to track the process frequency response and update the controller for the perturbed process. Load disturbance is often of low frequency and step signals are theoretically used as prototype disturbance (Astrom and Hagglund, 1995). Hang *et. al.* (1995) have discussed the adaptation from step load disturbance response. However, in practice a load disturbance is seldom a pure step signal. In a new adaptive method proposed in this section, the load disturbance is not limited to a step signal, it can be a signal that is generated by inputting a step or an impulse through an unknown rational function dynamics and thus the new method can work with quite a large range of load disturbances.

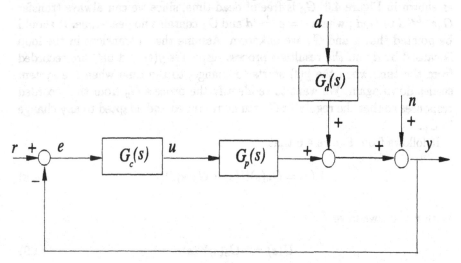

FIGURE 4.6. Control system.

4.3.1   Process Estimation from Load Disturbance Response

Consider a system shown in Figure 4.6, where $G_c(s)$ is a controller and $G_p(s)$ is a process. d and n are load disturbance and measurement noise respectively. The reason why the system configuration in Figure 4.6 is adopted instead of the FSA one is that the new method to be described below employs only information on $u(t)$ and $y(t)$ to update process estimation and is thus general and applicable to different types of control schemes such as single loop control, internal model control and FSA schemes, as will be seen in the subsequent two subsections. And Figure 4.6 enables us to deal with one element G_c only and thus greatly simplifies our presentation. One may regard G_c as an equivalent single-loop controller if a control scheme other than single-loop one is actually in use.

Suppose that the system has been in its steady state. A process transient may be caused by set-point change or load disturbance. In the case of set-point change, normal process estimation methods can be employed to re-establish the process model. However, in industrial practice, set-point is often kept constant. If no set-point change has been made, any significant transient must be the result of some load disturbance. Load disturbance is often encountered in process control and the adaptation from load disturbance is thus attractive. However, an inherent property of disturbance is that they cannot be predicted exactly, they are usually modeled in some prototype disturbances: impulse, step, ramp, sinusoid (Astrom and Hagglund, 1995). The impulse and step like disturbances are commonly encountered in practice and we focus on these two kinds of load disturbances.

We can always collect all these disturbances as an equivalent disturbance d acting at the process output y through an unknown dynamic element G_d, as shown in Figure 4.6. G_d is free of dead-time, since we can always transfer $G_d e^{-L_d s} d$ to $G_d \bar{d}$, where $\bar{d} = e^{-L_d s} d$ and G_d contains no dead-time. It should be pointed that d and G_d are unknown. Assume that a transient in the loop is caused by d and the resultant process responses $y(t)$ and $u(t)$ are recorded from the time when the $y(t)$ starts to change, to the time when the system settles down again. We want to re-identify the process G_p from the recorded responses so that the regulator G_c can be re- tuned and adapted to any change in G_p.

It follows from Figure 4.6 that

$$Y(s) = G_p(s)U(s) + G_d(s)D(s). \qquad (4.8)$$

With $r = 0$, we have

$$U(s) = -G_c(s)Y(s). \qquad (4.9)$$

and

$$\frac{Y(s)}{D(s)} = \frac{G_d(s)}{1 + G_p(s)G_c(s)}. \qquad (4.10)$$

Since d is unmeasurable, we check the new steady state of the process input, and can then infer

$$D(s) = \begin{cases} 1, & if\ u(\infty) = u(0); \\ \frac{1}{s}, & if\ u(\infty) \neq u(0). \end{cases} \qquad (4.11)$$

For $Y(s)$ in (4.10), we can calculate $Y(s)$ at $s = j\omega$ as follows. Suppose, without loss of generality, that the system is in the zero operating point before the load disturbance occurs and the controller contains integrator. $y(t)$ will return to its initial value after transient and $Y(j\omega)$ can be computed using Fast Fourier Transformation (FFT) of $y(t)$ (Hang et. al., 1995). With $Y(j\omega)$ calculated and $D(j\omega)$ given in (4.11), can $G_p(j\omega)$ and $G_d(j\omega)$ be uniquely determined from (4.10)? Let us look at the following example.

Example 4.5:   Suppose $G_c(s) = 1$, $G_p(s) = \frac{1}{s+2}$, $G_d(s) = \frac{1}{s+4}$, Equation (4.10) becomes

$$\frac{Y(s)}{D(s)} = \frac{s+2}{(s+3)(s+4)}.$$

Suppose $G_c(s) = 1$, $G_p = \frac{2}{s+2}$, $G_d = \frac{1}{s+3}$, Equation (4.10) also gives

$$\frac{Y(s)}{D(s)} = \frac{s+2}{(s+3)(s+4)}.$$

The example means that though we may have the same $Y(s)$ and $D(s)$, we cannot uniquely determine $G_p(s)$ and $G_d(s)$ at the same time. We find that under some conditions, $G_p(s)$ and $G_d(s)$ can be solitarily determined.

Lemma 4.1:   If a process $G_p(s)$ has some dead-time while $G_d(s)$ not(this is true), then with $Y(s)$ and $D(s)$ known, $G_p(s)$ and $G_d(s)$ can be uniquely determined.

PROOF: Suppose that besides $G_p(s) = G_{p0}e^{-Ls}$ and $G_d(s)$, $\hat{G}_p(s) = \hat{G}_{p0}(s)e^{-\hat{L}s}$ and $\hat{G}_d(s)$ also satisfy (4.10). From (4.10), we have

$$\frac{1}{G_d(s)} + \frac{G_c(s)}{G_d(s)}G_{p0}(s)e^{-Ls} = \frac{1}{\hat{G}_d(s)} + \frac{G_c(s)}{\hat{G}_d(s)}\hat{G}_{p0}(s)e^{-\hat{L}s}. \qquad (4.12)$$

The first items on both hand-sides of (4.12) are dead-time free parts while the 2nd items contain dead-time. To ensure that at all $s = j\omega$, $\omega \in R$, (4.12) is valid, the dead-time free parts on both hand-sides should be equal. So will the parts containing dead-time. Thus, one sees that

$$\begin{cases} G_d(s) = \hat{G}_d(s), \\ G_{p0}(s) = \hat{G}_{p0}(s), \\ L = \hat{L}, \end{cases}$$

hence the result.

It is fortunate that a large number of processes in process control contain some dead-time and we can use $Y(s)$ and $D(s)$ to calculate $G_p(s)$ and $G_d(s)$ simultaneously. It is known (Hang, 1991; Halevi, 1991) that most industrial processes can be approximately described by a low order plus dead-time model as

$$G_p(s) = \frac{\beta s + 1}{\alpha_1 s^2 + \alpha_2 s + \alpha_3}e^{-Ls}. \qquad (4.13)$$

G_d is modeled as

$$G_d(s) = \frac{\gamma s + 1}{\lambda_1 s^2 + \lambda_2 s + \lambda_3}. \tag{4.14}$$

which can represent both the monotonic and oscillatory load disturbance responses. It follows from (4.8) that

$$\frac{\beta s + 1}{\alpha_1 s^2 + \alpha_2 s + \alpha_3}e^{-Ls}U(s) + \frac{\gamma s + 1}{\lambda_1 s^2 + \lambda_2 s + \lambda_3}D(s) = Y(s). \tag{4.15}$$

Estimating the parameters in (4.15) is not an easy job. However, (4.15) can be transfered (Whitfield, 1986) to

$$a_1 s^4 Y(s) + a_2 s^3 Y(s) + \cdots + \lambda_3 Y(s)$$
$$= b_1 s^3 U(s)e^{-Ls} + \cdots + b_3 s U(s)e^{-Ls} + \frac{\lambda_3}{\alpha_3}U(s)e^{-Ls} \tag{4.16}$$
$$+ c_1 s^3 D(s) + \cdots + c_3 s D(s) + D(s).$$

At $s = j\omega$, (4.16) is a linear equation in term of $[a_1, \ldots, a_4, \lambda_3, b_1, b_2, b_3, \frac{\lambda_3}{\alpha_3}, c_1, c_2, c_3]$, where $Y(j\omega)$ and $D(j\omega)$ are known and $U(j\omega)$ in (4.16) is calculated using the method in Hang *et. al.* (1995), which is shown below. The time response of process input $u(t)$ is decomposed into

$$u(t) = u(\infty) + \Delta u(t), \tag{4.17}$$

where $u(\infty)$ is the steady state, and $\Delta u(t)$ is the transient response. It follows that the process frequency response is

$$U(j\omega) = \frac{u(\infty)}{j\omega} + \Delta U(j\omega), \tag{4.18}$$

where $\Delta U(j\omega)$ can be estimated using Fast Fourier Transformation (FFT) of $\Delta u(t)$. Then, equation (4.16) is solved with Least Squares method. Equation (4.16) can be further simplified, since $\frac{\lambda_3}{\alpha_3}$ in (4.16) can be directly calculated from the known $u(t)$, $y(t)$ and the assuming $d(t)$.

Lemma 4.2: If the system is stable, $G_c(s)$ has an integrator, and d is a step disturbance, then $\frac{\lambda_3}{\alpha_3} = -\frac{1}{u(\infty)}$.

PROOF: Applying the Final Value Theorem to (4.15) gives

$$\frac{1}{\alpha_3}u(\infty) + \frac{1}{\lambda_3}d(\infty) = y(\infty).$$

Due to the integrator in the controller $G_c(s)$, $y(\infty) = 0$, and $d(\infty)$ is assumed to be 1 by (4.11). So $\frac{\lambda_3}{\alpha_3} = -\frac{1}{u(\infty)}$.

Lemma 4.3:   If the system is stable, $G_c(s)$ has an integrator, and d is an impulse disturbance, then $\frac{\lambda_3}{\alpha_3} = -\frac{1}{U(0)}$.

PROOF: Substituting $s = 0$ into (4.10) yields

$$\frac{Y(0)}{D(0)} = \frac{G_d(0)}{1 + G_p(0)G_c(0)}. \tag{4.19}$$

where $D(0) = 1$, $G_d(0) = \frac{1}{\lambda_3}$, $G_p(0) = \frac{1}{\alpha_3}$ and $G_c(0) = \infty$ due to the integral in controller $G_c(s)$. It is obvious from (4.19) that $Y(0) = 0$. With this result, substituting $s = 0$ into (4.15) gives us

$$\frac{\lambda_3}{\alpha_3} = -\frac{1}{U(0)}.$$

Lemma 4.2 and Lemma 4.3 reduce the parametric estimation work. Equation (4.16) can be rearranged into

$$a_1 s^4 Y(s) + a_2 s^3 Y(s) + \cdots + a_5 Y(s)$$
$$= b_1 s^3 U(s)e^{-Ls} + \cdots + b_3 sU(s)e^{-Ls} \tag{4.20}$$
$$+ c_1 s^3 D(s) + \cdots + c_3 sD(s) + [D(s) + \frac{\lambda_3}{\alpha_3}U(s)e^{-Ls}].$$

where only 11 coefficients need to be determined.
    Equation (4.20) is re-written as

$$\Phi X = \Gamma, \tag{4.21}$$

where $\Phi = [s^4 Y(s), \ s^3 Y(s), \cdots, Y(s), \ -s^3 U(s)e^{-Ls}, \cdots, \ -sU(s)e^{-Ls}, \ -s^3 D(s), \cdots, -sD(s)]$, $\Gamma = D(s) + \frac{\lambda_3}{\alpha_3}U(s)e^{-Ls}$ and $X = [a_1, a_2, \ldots, a_5, b_1, b_2, b_3, c_1, c_2, c_3]^T$ are the real parameters to be estimated. Assume first that the process dead-time L is known, then with frequency responses $Y(j\omega_i)$, $U(j\omega_i)$ and $D(j\omega_i)$, $i = 1, 2, \cdots, m$, computed, equation (4.21) yields a system of linear algebraic equations at $s = j\omega_i$, $i = 1, 2 \cdots, m$. We can obtain the least square solution X in (4.21). The process frequency response $G_p(j\omega)$ can then be computed by

$$G_p(j\omega) = \frac{b_1 s^3 + b_2 s^2 + b_3 s + \frac{\lambda_3}{\alpha_3}}{a_1 s^4 Y + a_2 s^3 + a_3 s^2 + a_4 s + a_5} e^{-Ls}. \tag{4.22}$$

This solution in fact depends on L if L is unknown. The fitting error for (4.21) is given by

$$J(L) = \|\Phi(\Phi^T \Phi)^{-1} \Phi^T \Gamma - \Gamma\|_2, \tag{4.23}$$

which is a scalar nonlinear algebraic equation in one unknown L only. The error is then minimized with respect to L in the given interval, which is an iterative problem on one parameter L. Each iteration needs to solve a Least Squares problem corresponding to a particular value of L. The model parameters are determined when the minimum J is reached. To facilitate the solution further, we next derive some bounds for L so that the search can be constrained to a small interval. This will greatly reduce computations, improve numerical property, and produce a unique solution. It is noted that the phase lag contributed by the rational part of the model, $G_{p0}(s) = \frac{\beta s + 1}{a_1 s^2 + a_2 s + a_3}$, is bounded as

$$\arg G_{p0}(j\omega) \in [-\pi, \frac{\pi}{2}], \quad \forall \omega \in (0, \infty), \tag{4.24}$$

so that we can impose an upper bound $\bar{L}$ and a lower bound $\underline{L}$ on L:

$$\bar{L} = \min \left\{ -\frac{\arg G_p(j\omega_k) - \frac{\pi}{2}}{\omega_k} \right\}, \quad k = 1, 2, \cdots, m, \tag{4.25}$$

and

$$\underline{L} = \max \left\{ -\frac{\arg G_p(j\omega_k) + \pi}{\omega_k} \right\}, \quad k = 1, 2, \cdots, m. \tag{4.26}$$

Actually, from a relay feedback or set-point change, one can directly find out a gross estimate for dead time by measuring the time $\hat{L}$ between the control signal change to the output starting to move. Another possible bound may then be

$$L \in [0.5\hat{L} \quad 1.5\hat{L}]. \tag{4.27}$$

Extensive simulation and real-time experiments show that within a reasonable bound, (4.23) exhibits a concave relationship with respect to L and yields a unique solution. Once L is determined, X is computed from (4.21).

Remark 4.1   If the process dead time L is unchanged since the last identification of G_p , no iteration is needed to solve (4.21). This is a special case that the bound for L is specified as a zero interval. This case greatly simplifies the identification. It may be true in many practical cases as process dynamics perturbations are usually associated with operating point changes and/or load disturbance, which mainly cause time constant/gain changes. Furthermore, any small dead time change can be discounted in other parameter changes.

Remark 4.2   Weighting can be considered when solving (4.21). Distortions caused by transferring a nonlinear problem of (4.15) to a linearized problem of (4.21) can be reduced using weighting (Whitfield, 1986). To have a better result, more weights should be given to the frequency region where $Y(j\omega)$ and $U(j\omega_i)$ has less computation error. This is especially important when real-time data is considered.

Remark 4.3   The measurement noise n in Figure 4.6 is usually of high frequency while the process frequency response of interest for control analysis and design is usually in the low frequency region. In particular, the process frequency response from 0 to the critical frequency ω_π is mostly critical for Controller Design. We found that in our experiments the measurement noise is indeed in the fairly high frequency region. Therefore, a low pass filter can be employed to reduce the measurement noise. The cut-off frequency of the filter is determined with respect to the process frequency region of interest. A possible choice is $(3 \sim 5)\omega_\pi$.

Remark 4.4   If the set-point is not changed after the system enters its steady state, the transients in process input and output may also be caused by the process static gain change. The process response under this change is equivalent to the process response under a load disturbance. If the process gain abruptly changes to its final value, then, our proposed process estimation from load disturbance response is still valid. This can be shown by Figure 4.7. Suppose that at $t = t_0^-$, the systems in Figure 4.7(a) and Figure 4.7(b) have both entered the steady states. At $t = t_0$, the process gain in Figure 4.7(a) suddenly jumps from 1 to k. We have $\nu(t_0^-) = u_0$, $\nu(t_0^+) = ku_0$ and $y(t_0^-) = y_0$, where u_0 and y_0 are the steady states. The system in Figure 4.7(b) before $t = t_0$ is also in steady state with $d = -(k-1)u_0$. At $t = t_0$ and afterwards, the load disturbance changes to zero. From Figure 4.7(b), we also have $\nu(t_0^-) = u_0$, $\nu(t_0^+) = ku_0$ and $y(t_0^-) = y_0$. From $t = t_0$ onwards, the systems in Fig 4.7(a) and Figure 4.7(b) have the same structure and parameters, and the initial values at each point in the two plots are the same. Therefore, the systems in Figure 4.7(a) and Figure 4.7(b) have the same responses after $t = t_0$ and they are equivalent. Figure 4.7(b) can be further modified to Figure 4.6, and our estimation method can be applied. This is demonstrated by an example later on.

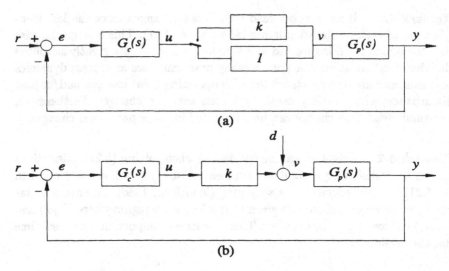

FIGURE 4.7. Equivalent systems when gain changes.

4.3.2 PID Adaptation

After the process frequency response $G_p(j\omega_i)$, $i = 1, 2, \cdots, m$, has been re-identified, a PID controller:

$$G_c(s) = K_p \left(1 + \frac{1}{T_i s} + T_d s \right),$$

may be re-tuned by matching $G_p(j\omega_i)G_c(j\omega_i)$ to the desired open-loop $G_{yr}(j\omega_i)$. The desired closed-loop transfer function G_{yr}^* is set (Wang $et.$ $al.$, 1997) as

$$G_{yr}^*(s) = \frac{\omega_n^2}{s^2 + 2\zeta\omega_n s + \omega_n^2} e^{-Ls}$$

according to the given specifications. If the control specifications are not given by the user, then the default settings can be used for the parameter ζ and $\omega_n L$, and preferably they are $\zeta = 0.707$ and $\omega_n L = 2$, which implies that the overshoot of the objective set-point step response is about 5%, the phase margin is 60° and the gain margin is 2.2. The desired open-loop $G_{yr}(j\omega)$ is thus given by

$$G_{yr}(j\omega_i) = \frac{G_{yr}^*(j\omega_i)}{1 - G_{yr}^*(j\omega_i)}.$$

The PID controller parameters can be obtained with linear least squares method (Wang, et. al., 1997). The controller is then able to adapt to the possible changes in process.

Several typical processes are employed in demonstrating the proposed method. For assessment of accuracy, the identification error is here measured by worst case error

$$ERR = \max_i \left\{ \left| \frac{\hat{G}_p(j\omega_i) - G_p(j\omega_i)}{G_p(j\omega_i)} \right| \times 100\%, i = 1, 2, \ldots, M \right\}, \qquad (4.28)$$

where $G_p(j\omega_i)$ and $\hat{G}_p(j\omega_i)$ are the actual and the estimated process frequency responses respectively. The Nyquist curve for phase ranging from 0 to $-\pi$ is being considered since this part is most significant for control design.

Example 4.6   Consider the control system in Figure 4.6. Suppose initially that the system is in the steady state and for some reason the process is changed to

$$G_p(s) = \frac{1}{s^2 + 3s + 2} e^{-s}.$$

Thereafter, a step disturbance comes into the system through an unknown dynamics:

$$G_d(s) = \frac{1}{s + 3}.$$

And the significant transients in the system are detected. The process output $y(t)$ and input $u(t)$ are then logged until the process settles down again. $Y(j\omega)$ is computed using FFT, $U(j\omega)$ is determined using (4.18) while $D(j\omega)$ is taken as step-type since the steady state of $u(t)$ reaches a new value. The process frequency response is calculated by (4.21) and (4.22), which is shown in Figure 4.8. The ERR is 0.47%. This indicates that the proposed method provides a very accurate process frequency response.

A step disturbance may come through different unknown dynamics. Suppose that they are respectively $G_d(s) = \frac{1}{s^2 + 4s + 3}$ and $G_d(s) = \frac{1}{s^2 + 5s + 10}$ and the identification errors are listed in Table 4.1. An impulse input is also considered, the re-identification results are also very satisfactory, as shown in Table 4.1.

Example 4.7   Suppose that a second order oscillatory process has perturbed to

$$G_p(s) = \frac{1}{s^2 + s + 1} e^{-2s}.$$

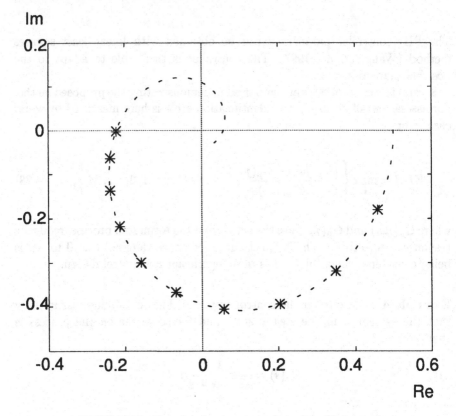

FIGURE 4.8. Nyquist plot. $\cdots + \cdots$ Actual, $\times$ Estimated

From the step/pulse-like load disturbance responses, the process can be re-estimated. Different load disturbances generated by passing a step or an impulse through different unknown dynamics are employed to illustrate the proposed method. The estimation errors are listed in Table 4.1.

Example 4.8 For a high order process:

$$G_p(s) = \frac{1}{(s+1)^5} e^{-4s},$$

it is re-identified under different load disturbance cases. The estimation errors ERR are given in Table 4.1.

The above examples are of noise free cases. The estimation results under noise pollution are also studied, where the noise source is a white noise. The noise-to-signal ratio (Haykin, 1989) is defined as Noise-to-Signal Power Spectrum Ratio N_1. The low pass filter is selected as a Butterworth low pass filter whose cut-off frequency is chosen as $3 \sim 5\omega_\pi$. The process in Example 4.8 is used to illustrate the results under different noise levels, where the unknown disturbance is obtained by passing a step signal through $\frac{1}{s^2+4s+3}$. The estima-

Table 4.1. Identification Error from Load Disturbance

$G_p(s)$	$\frac{1}{s^2+3s+2}e^{-s}$		$\frac{1}{s^2+s+1}e^{-2s}$		$\frac{1}{(s+1)^5}e^{-4s}$	
$G_d(s)$	Step	Pulse	Step	Pulse	Step	Pulse
$\frac{1}{s+3}$	0.47%	0.13%	3.85%	0.72%	1.33%	0.44%
$\frac{1}{s^2+4s+3}$	3.30%	1.21%	5.51%	0.59%	1.24%	0.95%
$\frac{3}{s^2+5s+10}$	1.40%	0.75%	0.29%	0.51%	3.86%	1.37%

tion errors are shown in Table 4.2, which are quite satisfactory.

Table 4.2. Estimation Error under Different Noise Levels

N_1	0	4.50%	9.00%	17.5%	25.5%
ERR	1.24%	3.11%	5.91%	8.64%	12.33%

To test the robustness of the method, we apply a non-step/impulse input to the unknown element $G_d(s)$. For a process:

$$G_p(s) = \frac{1}{s^2 + 3s + 2}e^{-s}.$$

a non-step disturbance shown in Figure 4.9(a) is introduced to the system through $G_d(s) = \frac{1}{s+3}$. We still treat the non-step input as a step-input and perform the same re-identification procedure. The resultant error ERR is 9.82%. A non-impulse input as in Figure 4.9(b) is also employed to show the robustness of the proposed method. The process frequency response estimation error ERR is 9.33%.

Real-Time Test   The method was tested on the Dual Process Simulator KI 100. To observe the noise effect in the test, apart from the inherent test environment noise, extra noise with peak-to-peak value of 0.1 V is introduced using the noise source in the Simulator.

As a demonstration, we choose the process from the Simulator as

$$G_p(s) = \frac{1}{(5s + 1)^2}e^{-5s}.$$

With our PID design method (Wang, et. al., 1997), the PID controller is obtained as

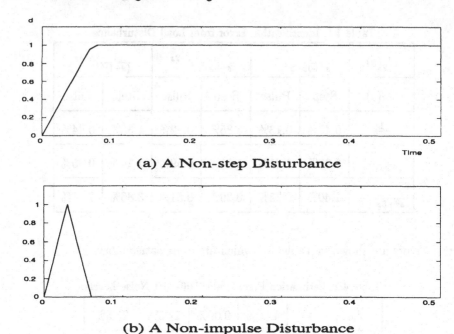

(a) A Non-step Disturbance

(b) A Non-impulse Disturbance

FIGURE 4.9. Disturbance signal for robustness testing

$$G_c(s) = 0.66 \left(1 + \frac{1}{10.04s}\right). \tag{4.29}$$

The process response to a set-point change at $t = 0$ is pretty good, which is shown in Figure 4.10. Suppose that after the process settles down, the process gain suddenly changes from 1 to 2 at $t = 66s$ as

$$G_p(s) = \frac{2}{(5s + 1)^2} e^{-5s}.$$

The transients of the process input and output are recorded as shown in Figure 4.10. As we have shown before, the process response under gain change is equivalent to a step load disturbance. Thus, we still can re-estimate the process model using the proposed method in Section 4.3.1 and the PID controller is re-tuned accordingly. This results in a new controller:

$$G_c(s) = 0.45 \left(1 + \frac{1}{11.67s} + 0.11s\right). \tag{4.30}$$

The next step set-point change at $t = 228s$ leads to a quite satisfactory response, as shown in Figure 4.10. For comparison, if the controller in (4.29) was still used for the new process, the resultant response (dashed line) is also displayed in Figure 4.10. Assume next that the process is further perturbed from 2nd order to 4th order and dead-time changes to 2.5 as

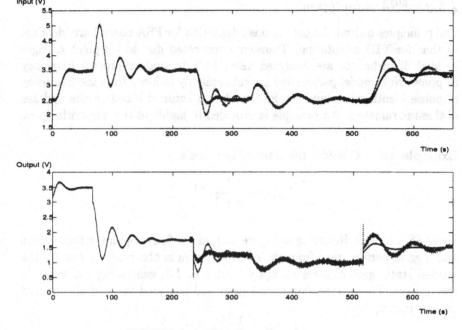

FIGURE 4.10. Real-time adaptation test.

$$G_p(s) = \frac{2}{(5s+1)^4}e^{-2.5s}.$$

An unknown load disturbance occurs, which is generated by applying a step signal through an unknown disturbance channel:

$$G_d = \frac{1}{(2s+1)^2}.$$

We utilize the load disturbance response to run the adaptation scheme. The process is updated and the controller is adjusted to

$$G_c(s) = 0.33\left(1 + \frac{1}{15.23s}\right). \tag{4.31}$$

to adapt to the process perturbation. A set-point change occurs at $t = 510s$. The control performance of the new process under controller in (4.30) is shown in Figure 4.10 and the adaptation performance by our method is also illustrated there. The real-time test shows that our method is capable of adapting the controller using a benign unknown load disturbance response.

4.3.3 FSA Adaptation

The principles behind the continuous adaptation for FSA control are identical to that for PID adaptation. Transients generated due to set-point changes or load disturbances are analyzed using FFT to update process frequency response. New model parameters are subsequently induced from the frequency response identification and the FSA control is retuned based on the updates of these parameters. An example is provided to highlight the main principles.

Example 4.9 Consider the time-delay process:

$$G_p(s) = \frac{1}{s+1}e^{-4s},$$

under FSA control. Referring to Figure 4.11, at $t = 0$, a set-point change occurs and a good performance from the control system is observed. At $t = 15$, the process static gain changes abruptly from 1 to 1.3, generating transients in the process. These transients were logged and analyzed to yield an updated process model,

$$G_p(s) = \frac{1.28}{s+1}e^{-3.9s},$$

and the FSA control is adjusted accordingly. The next set-point change at $t = 50$ shows the process is back under tight control again.

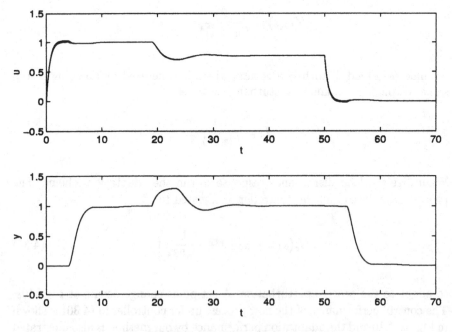

FIGURE 4.11. FSA adaptation

4.3.4   Conclusion

A new adaptation control scheme based on unknown load disturbance response has been developed. It extends the method of controller adaptation from a pure step load disturbance. The proposed method can update a controller using unknown step/pulse-like load disturbance response and thus widens the scope of application. This method is especially useful when the process parameters change significantly. The effectiveness of the method in process re-modelling and controller re-design has been demonstrated through our simulations and real-time tests.

4.4   Unstable Processes

Consider an unstable process characterized by a first-order plus dead time model:

$$G_p(s) = \frac{K}{Ts - 1}e^{-Ls}, \tag{4.32}$$

Then, the generalized process is

$$G(s) = \frac{a_o}{s(s - b_1)}e^{-Ls}, \tag{4.33}$$

where $a_o = -\frac{K}{T}$ and $b_1 = \frac{1}{T}$. Since it is a second-order system, $p(s)$ is chosen as a degree 2 polynomial

$$p(s) = s^2 + 2\zeta\omega_o s + \omega_o^2, \tag{4.34}$$

where ω_o and ζ are the natural frequency and the damping factor of the desired closed-loop response, respectively. In order to simplify design, the observer polynomial $q(s)$, $k(s)$ and $h(s)$ are chosen as

$$q(s) = s + \frac{b_1}{\alpha} = s + \frac{1}{\alpha T}, \quad \alpha > 0, \tag{4.35}$$
$$k(s) = k, \tag{4.36}$$
$$h(s) = h_1 s + h_0, \tag{4.37}$$

where α is a user-specified parameter which relates to the location of the observer's zero. The polynomial equation (2.9) of Chapter 2 then reduces to

$$k(s)s(s - b_1) + h(s)a_o = (s + b_1)f_L(s), \tag{4.38}$$

which is very easy to solve. For the given $b(s) = s(s - b_1)$, and $p(s)$ in (4.34), Equations (2.6)-(2.8) gives

$$f_L(s) = f_1 s + f_0 = (c_1 + c_2 e^{b_1 L})s - b_1 c_1,$$

where $c_1 = \frac{\omega_o^2}{b_1}$ and $c_2 = -\frac{b_1^2 + 2b_1 \omega_o \zeta + \omega_o^2}{b_1}$. Substituting them into (4.38) yields the solution

$$p(s) = s^2 + 2\zeta\omega_o s + \omega_o^2, \tag{4.39}$$

$$q(s) = s + \frac{1}{\alpha T}, \tag{4.40}$$

$$k(s) = T\omega_o^2 - \left(\frac{1}{T} + 2\zeta\omega_o + \omega_o^2 T\right)e^{\frac{L}{T}}, \tag{4.41}$$

$$h(s) = \frac{(-(\alpha+1)k(s) + \alpha T\omega_o^2)s + \omega_o^2}{\alpha K}. \tag{4.42}$$

All the required polynomials $p(s)$, $q(s)$, $k(s)$ and $h(s)$ to implement the FSA controller are now available from (4.39)-(4.42), respectively. The ideal closed-loop transfer function can be shown to be

$$G_{yr}^*(s) = \frac{(-(\alpha+1)k(s) + \alpha T\omega_o^2)s + \omega_o^2}{(s^2 + 2\zeta\omega_o s + \omega_o^2)(\alpha T s + 1)}e^{-sL}.$$

In this particular design, it is not possible to cancel the closed-loop left-hand-plane zero contributed by the polynomial $h(s)$ with the poles from polynomials $p(s)$ and $q(s)$. This zero typically gives a sharp resonance to the frequency response of $G_{yr}^*(j\omega)$ which translates to a large overshoot in the set-point response of the system, particularly for dead time dominant processes (because of the $e^{L/T}$ term in $k(s)$ bringing the zero close to the origin). As in the case for stable system, ζ, ω_o and α are parameters relating to the closed-loop performance which are specified by the user.

For auto-tuning, both the critical point and an estimate of the process time-delay are necessary to invoke the modelling algorithm of Section 3.6.1. The critical point is obtained from the normal or enhanced relay experiment. The time-delay estimate may be obtained from an observation of the time propagation of the input and output signals of the process, as the time interval between a relay switch and the following extremum point in the process variable (Astrom 1988b). A simulation example is provided here to illustrate the auto-tuning principles.

Example 4.9   Consider a heat transfer process in a non-isothermal continuous stirred-tank reactor (CSTR) described by

$$G_p(s) = \frac{1}{10s - 1}e^{-2s},$$

where the output is the process temperature, and the input is the inflow of water to the cooling jacket. From $t = 0$ to $t = 28$, the process is put under relay feedback, and from the resultant limit cycle oscillation, the critical point is obtained as $(k_u, \omega_u) = (5.75, 0.695)$. The time-delay is estimated as $L = 2$. By the algorithm in Section 3.6.1, the model is identified as

$$\hat{G}_p(s) = \frac{0.9}{8s - 1}e^{-2s}.$$

The desired closed-loop polynomial is specified as

$$p(s) = s^2 + 20s + 100,$$

and the observer polynomial is specified with $\alpha = 0.1$. The following polynomials of the FSA system are then obtained :

$$
\begin{aligned}
h(s) &= 3982s - 1111, \\
k(s) &= -253, \\
q(s) &= s + 1.25.
\end{aligned}
$$

The FSA system is commissioned at $t = 28$. At $t = 35$, a set-point change occurs, and at $t = 53$, a 10% step disturbance occurs. The auto-tuning and the subsequent closed-loop performance is shown in Figure 4.12.

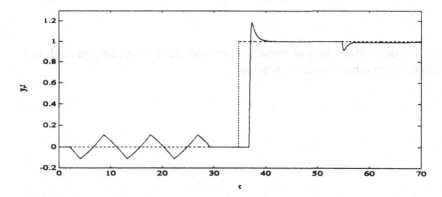

FIGURE 4.12. FSA auto-tuning and closed-loop performance – Example 4.9.

4.5 Problems

1. Consider a process described by the following transfer function :

$$G(s) = \frac{Y(s)}{U(s)} = \frac{5}{(3s+1)^2} e^{-s}.$$

The desired closed-loop response parameters are $\zeta = 1$ and $\omega_0 = 10$. Design a FSA control system based on a first-order model approximation of the actual process.

2. Consider an unstable process described by the following transfer function :

$$G(s) = \frac{Y(s)}{U(s)} = \frac{2}{(10s-1)^2} e^{-0.5s}.$$

Write a simulation program to demonstrate the applicability of FSA control to stabilise the process. Choose and justify the desired closed-loop parameters.

3. Why is auto-tuning useful? Survey the types of auto-tuning functions present in the industrial controllers available.

4. Why is continual self-tuning from load disturbance responses so important? How does it compare with the self-tuning regulator?

5. Write a simulation program to implement the algorithm in Section 4.3 and perform the simulation for a process of your choice.

6. What difficulties do you expect to extend the auto-tuning procedure in Section 4.2 to a multivariable process?

Robustness Analysis and Design

In the FSA algorithms development of Chapter 2, it has been assumed that the nominal process (model) used for design matches the actual process perfectly, i.e. there exist no modelling errors. In practical situations, a model can never be a perfect representation of the actual process, although it can be valid for the purposes of simulation and control. In certain cases, this may cause a nominally stable system to become unstable under the slightest perturbation to the closed-loop system. Under modelling uncertainty, it is therefore essential and necessary to analyze the robustness of the FSA system with respect to stability and performance.

5.1 Practical Stability

Assume that $\hat{G}(s) = \frac{\hat{a}(s)}{\hat{b}(s)}e^{-\hat{L}s}$ is a model for the actual generalized process $G(s) = \frac{a(s)}{b(s)}e^{-Ls}$. $G(s)$ is unknown to us and only $\hat{G}(s)$ is available for FSA system design. In general, $\hat{G}(s) \neq G(s)$. For any practical use, a FSA system must remain stable for small difference between $G(s)$ and $\hat{G}(s)$. To address this problem formally, following (2.5)–(2.9) of Chapter 2 using the model $\hat{G}(s)$ instead of $G(s)$, it can be shown that the closed-loop transfer function is

$$Y(s) = -\frac{h(s)a(s)}{b(s)w(s)}e^{-sL}R(s) \tag{5.1}$$

where

$$w(s) := \frac{q(s)p(s)}{\hat{b}(s)} + \frac{h(s)\hat{a}(s)}{\hat{b}(s)}e^{-\hat{L}s} - \frac{h(s)a(s)}{b(s)}e^{-Ls}. \tag{5.2}$$

Denoting $\bar{w}(s) = b(s)w(s)$, the characteristic equation is thus given by

$$\bar{w}(s) = 0.$$

The system stability depends entirely on the location of the zeros of the characteristic equation. Let $\sum$ denote the set of real parts of all the zeros of $\bar{w}(s)$:

$$\sum := \{\sigma | \sigma = \mathrm{Re}z, \bar{w}(z) = 0\}.$$

Noting that $\sum$ is in general an infinite set (Bellman and Cooke, 1963), stability of the system is then defined as follows.

Definition 5.1   The closed-loop system is said to be stable iff $\sum$ has a negative upper bound.

The nominal system with $G(s) = \hat{G}(s)$ is always designed such that it is stable. But if there is a perturbation in the process, the system will not necessarily preserve its stability. This is shown in the following example.

Example 5.1   Consider a process with the transfer function:

$$G(s) = \frac{\epsilon s^2 + \epsilon s + 1}{(\epsilon s + 1)(s + 1)} e^{-s},$$

where $\epsilon > 0$ is a small parameter. It is a common practice to obtain the nominal process by neglecting the small parameter ϵ as

$$\hat{G}(s) = \frac{1}{s+1} e^{-s}.$$

Solving the design equations (2.6)-(2.9) with stable $p(s)$ and $q(s)$ specified as

$$p(s) = s + \alpha, \quad \alpha > \epsilon + 1 \tag{5.3}$$

and

$$q(s) = 1$$

gives

$$f(s) = 1 - \alpha, \quad \frac{f(s)}{\hat{b}(s)} = \frac{1 - \alpha}{s + 1},$$

$$f_L(s) = (1 - \alpha)e^{-1}, \quad k(s) = 0,$$

and

$$h(s) = (1 - \alpha)e^{-1}.$$

In this example, $p(s)$, $q(s)$ and $\hat{b}(s)$ are all stable, and $w(s)$ has the same zeros in the closed right half complex plane as those of

$$\tilde{w}(s) := w(s)\frac{\hat{b}(s)}{p(s)q(s)} = 1 + \gamma(s)e^{-s},$$

where

$$\gamma(s) = \frac{\alpha - 1}{e} \cdot \frac{\epsilon s^2}{(\epsilon s + 1)(s + \alpha)}.$$

It can be shown (Bellman and Cooke, 1963) that the zeros of $\tilde{w}(s)$ make up an infinite chain asymptotic to those of the comparison function:

$$f_c(s) := 1 + \gamma(\infty)e^{-s}.$$

Since the chain of the zeros of $f_c(s)$ is located on the line Re $s = \log\gamma(\infty) = \log\frac{\alpha-1}{e} > 0$ by (5.3), $\tilde{w}(s)$ has an infinite number of zeros in the right-half plane no matter how small ϵ is. In other words, the closed-loop system is only stable when $\epsilon = 0$, and an infinitesimal perturbation of the process transfer function coefficients will destabilize a nominally stable system. Obviously, such a system is useless in industrial control and stability of the system in the face of mismatch has to be considered. The following definition is first made for practical stability.

Definition 5.2 Let a finite spectrum-assigned system with the nominal process $\hat{G}(s) = \frac{\hat{a}(s)}{\hat{b}(s)}e^{-\hat{L}s}$ be stable. The closed-loop system is said to be practically stable iff there exist positive numbers, ω_M and δ, such that the system is stable for each process $G(s) = \frac{a(s)}{b(s)}e^{-Ls}U(s)$ satisfying

$$\left|\frac{G(j\omega)}{\hat{G}(j\omega)} - 1\right| < \delta, \quad 0 \le \omega \le \omega_M. \tag{5.4}$$

Theorem 5.1 deals with strictly proper processes.

Theorem 5.1 *If $\frac{\hat{a}(s)}{\hat{b}(s)}$ is strictly proper and each $\frac{a(s)}{b(s)}$ is strictly proper and has the same number of unstable poles as $\frac{a(s)}{b(s)}$ has, then the closed-loop system is practically stable.*

PROOF Denoting $Z^+(\tilde{w})$ as the number of unstable zeros of the characteristic equation $\bar{w}(s) = 0$, it has to be shown that $Z^+(\bar{w}) = 0$ if the conditions of Theorem 5.1 hold. It is observed that

$$\bar{w}(s) = \frac{b(s)}{\hat{b}(s)} \cdot p(s)q(s)\tilde{w}(s), \tag{5.5}$$

where

$$\tilde{w}(s) := w(s)\frac{\hat{b}(s)}{q(s)p(s)} = 1 + \frac{\hat{b}(s)h(s)}{q(s)p(s)}\left[\frac{\hat{a}(s)}{\hat{b}(s)}e^{-\hat{L}s} - \frac{a(s)}{b(s)}e^{-Ls}\right].$$

By (5.2) and (5.5), $\bar{w}(s)$ has the following form:

$$\bar{w}(s) = b(s) \cdot \frac{\phi_o(s)}{\hat{b}(s)} - h(s)a(s)e^{-Ls}, \tag{5.6}$$

where

$$\phi_o(s) = q(s)p(s) + h(s)\hat{a}(s)e^{-\hat{L}s}.$$

It is now claimed that $\bar{w}(s)$ has no unstable poles, i.e.

$$P^+(\bar{w}) = 0, \tag{5.7}$$

($P^+(\bar{w})$ is the number of unstable poles of $\bar{w}$) since it can be inferred from (2.48) and (2.49) in the proof of Theorem 2.1 that $\phi_o(s)$ has exactly the same zeros as $\hat{b}(s)$ has.

Denoting by $N(k, \bar{w})$, the net number of clockwise encirclements of the point $(k, 0)$ by the image of the Nyquist D contour under $\bar{w}$, and by the principle of argument, it follows that

$$N(0, \bar{w}) = Z^+(\bar{w}) - P^+(\bar{w}).$$

Since $P^+(\bar{w}) = 0$, $Z^+(\bar{w}) = 0$ if and only if $N(0, \bar{w}) = 0$, i.e. the Nyquist curve of $\bar{w}(s)$ does not encircle the origin. Since $b(s)$ has the same number of unstable zeros as $\hat{b}(s)$ has, and $p(s)$ and $q(s)$ are user-specified and Hurwitz polynomials, (5.5) implies that

$$N(0, \bar{w}) = N(0, \tilde{w}) = N(-1, \tilde{w} - 1).$$

For stability, it is necessary that

$$N(-1, \tilde{w} - 1) = 0.$$

$\tilde{w}(s)$ can be rewritten as

$$\tilde{w}(s) = 1 + \frac{\hat{a}(s)h(s)}{q(s)p(s)} \cdot e^{-Ls} \cdot \left[1 - \frac{G(s)}{\hat{G}(s)}\right]. \tag{5.8}$$

or

$$\tilde{w}(s) - 1 = G^*_{yr}(s)l_m(s),$$

where $G^*_{yr}(s) = -\frac{\hat{a}(s)h(s)}{p(s)q(s)}e^{-s\hat{L}}$ is the ideal closed-loop transfer function between the set-point and process output.

Since $\frac{\hat{b}(s)}{p(s)}$ and $\frac{h(s)}{q(s)}$ are both proper, $\frac{\hat{a}(s)}{\hat{b}(s)}$ and $\frac{a(s)}{b(s)}$ are both strictly proper, it follows that for the part of the Nyquist D contour where $|s| \longrightarrow \infty$, and $\text{Re } s \geq 0$,

$$\left| G_{yr}^{*}(s) l_m(s) \right| = \left| \frac{\hat{b}(s) h(s)}{p(s) q(s)} (\hat{G}(s) - G(s)) \right|$$

$$\leq \left| \frac{\hat{b}(s)}{p(s)} \right| \cdot \left| \frac{h(s)}{q(s)} \right| \cdot \left[\left| \frac{\hat{a}(s)}{\hat{b}(s)} \right| \cdot \left| e^{-L s} \right| + \left| \frac{a(s)}{b(s)} \right| \cdot \left| e^{-L s} \right| \right] \longrightarrow 0. \qquad (5.9)$$

Hence, there exists a positive number ω_M such that

$$|\tilde{w}(j\omega) - 1| < 1, \quad \omega > \omega_M. \qquad (5.10)$$

The stability of $\frac{\hat{a}(s) h(s)}{q(s) p(s)} \cdot e^{-L s}$ implies that

$$\left| \frac{\hat{a}(j\omega) h(j\omega)}{q(j\omega) p(j\omega)} \cdot e^{-jL\omega} \right| < M, \quad 0 \leq \omega \leq \omega_M, \qquad (5.11)$$

for some positive M. Take a positive δ such that $M\delta < 1$, then for each process $G(s) = \frac{a(s)}{b(s)} e^{-L s}$ satisfying

$$\left| \frac{G(j\omega)}{\hat{G}(j\omega)} - 1 \right| < \delta, \quad 0 \leq \omega \leq \omega_M,$$

there holds

$$|\tilde{w}(j\omega) - 1| < 1, \quad 0 \leq \omega \leq \omega_M. \qquad (5.12)$$

Equations (5.10) and (5.12) together imply that

$$|\tilde{w}(s) - 1| < 1$$

for each point of the Nyquist Contour and thus the Nyquist curve of $\tilde{w}(s)$ does not encircle the origin and the proof is completed.

One notes that if, in addition to strict properness, $\frac{\hat{a}(s)}{\hat{b}(s)}$ and $\frac{a(s)}{b(s)}$ are stable, then the conditions of Theorem 5.1 are satisfied. The following corollary is obtained.

Corollary 5.1 *If $\frac{\hat{a}(s)}{\hat{b}(s)}$ and $\frac{a(s)}{b(s)}$ are stable and strictly proper, then the closed-loop system is practically stable.*

It may however happen that either $\frac{\hat{a}(s)}{\hat{b}(s)}$ or any $\frac{a(s)}{b(s)}$ is not strictly proper, then Theorem 5.1 does not apply. In this case, the following necessary and sufficient condition for practical stability can be established under certain assumptions on the processes.

Theorem 5.2   *If $\frac{\hat{a}(s)}{\hat{b}(s)}$ is stable and minimum phase, and all $\frac{a(s)}{b(s)}$ are stable, then the closed-loop system is practically stable iff*

$$\lim_{\omega \to \infty} (\kappa \omega^\lambda + 2)|G^*_{yr}(j\omega)| < 1, \qquad (5.13)$$

where the non-negative real κ and non-negative integer λ satisfy

$$\left| \frac{a(j\omega)/b(j\omega)}{\hat{a}(j\omega)/\hat{b}(j\omega)} - 1 \right| \leq \kappa \omega^\lambda, \quad \omega > \beta \geq 0.$$

PROOF   This result is a natural extension of one for Smith-Predictor Controller. Under the given conditions, Theorem 1 of Yamanaka and Shimemura (1987) is applicable to $\tilde{w}(s)$ and the result follows directly.

As the first application of Theorem 5.2, Example 5.1 is reconsidered. The model and design data yield

$$\left| \frac{a(j\omega)/b(j\omega)}{\hat{a}(j\omega)/\hat{b}(j\omega)} - 1 \right| = \left| \frac{\epsilon(j\omega)^2}{\epsilon j\omega + 1} \right| \leq \omega, \quad \omega > \beta \geq 0,$$

and

$$G^*_{yr}(s) = \frac{\alpha - 1}{e(s + \alpha)} e^{-s}.$$

It then follows that

$$\lim_{\omega \to \infty} (\omega + 2)|G^*_{yr}(j\omega)| = \frac{\alpha - 1}{e} > 1,$$

where the last inequality is due to (5.3). The condition (5.13) is thus violated. This verifies that the system is not practically stable.

One may note that in Example 5.1, $\frac{a(s)}{b(s)}$ is proper but not strictly proper and $\frac{\hat{a}(s)}{\hat{b}(s)}$ is strictly proper. It is interesting to consider the converse case too. Now, if $\frac{a(s)}{b(s)}$ is strictly proper and $\frac{\hat{a}(s)}{\hat{b}(s)}$ is only proper but not strictly proper (as in singular perturbation cases), then $\lambda = 0$ and $G^*_{yr}(\infty)$ is finite. It follows from (5.13) that whether or not the system is practically stable will depend on the value of $G^*_{yr}(\infty)$. In fact, one can get from (5.13), the upper bound for $|G^*_{yr}(\infty)|$ within which the system remains practically stable. In view of these discussions, it is concluded that practical instability may result if either the process or its model is not strictly proper.

5.2   Robust Stability

Practical stability deals with infinitesimal perturbation in the model and is the basic requirement of any practical control system. The developments in Section 5.1 show the FSA system will be practically stable if both process and model are strictly proper and contain the same number of unstable poles. However, practical stability is a weaker notion compared to the usual robust stability. For arbitrary magnitude in model perturbation, satisfaction of the practical stability condition is not sufficient and controller needs to be designed for robust stability. To this end, a key result is given in Theorem 5.3.

Theorem 5.3   *Assume that the family of processes $\prod$ with norm-bounded uncertainty is described by*

$$\prod = \left\{ G : \left| \frac{\hat{G}(j\omega) - G(j\omega)}{\hat{G}(j\omega)} \right| = |l_m(j\omega)| \leq \tilde{l}_m(\omega) \right\}, \qquad (5.14)$$

where $G(s) = \frac{a(s)}{b(s)} e^{-Ls}, \hat{G}(s) = \frac{\hat{a}(s)}{\hat{b}(s)} e^{-\hat{L}s}$, and $\tilde{l}_m(\omega)$ is the bound on the multi- plicative uncertainty $l_m(j\omega)$. Furthermore, both $\frac{a(s)}{b(s)}$ and $\frac{\hat{a}(s)}{\hat{b}(s)}$ are strictly proper and they have the same number of unstable poles. Then, the FSA system is robustly stable if

$$|G^*_{yr}(j\omega)| \, \tilde{l}_m(\omega) < 1 \;, \quad \forall \omega. \qquad (5.15)$$

PROOF.   Under the given conditions, the evaluation of (5.9) for the imaginary axis of the Nyquist Contour provides the robust stability condition, i.e.

$$|\tilde{w}(j\omega) - 1| < 1, \quad -\infty < \omega < \infty. \qquad (5.16)$$

Given $\tilde{w}$ in (5.8) and the specified uncertainty bound in (5.14), Theorem 5.3 follows directly.

Remark 5.1   For robust stability, the amplitude response of $|G^*_{yr}(j\omega)|$ should be shaped to lie below that of $\tilde{l}_m^{-1}(\omega)$. This can be done through a proper assignment of the ideal closed-loop poles via the desired closed-loop polynomial $p(s)$ and the observer polynomial $q(s)$. The closed-loop zeros contributed by the polynomials $\hat{a}(s)$ and $h(s)$, however, cannot be arbitrarily assigned. The dynamics introduced by these zeros also has to be accounted for in robust stability.

5.3   Performance Robustness

While being desirable in a practical environment where model uncertainty is an important issue, robust stability alone is not enough. Even if (5.15) is satisfied for the family $\prod$, there will exist a "worst case" process in $\prod$ for which the closed-loop system is on the verge of instability, and for which the performance is arbitrarily poor. Thus, it is necessary to ensure that some performance specifications are met for *all* processes in the family $\prod$. Performance specifications stated in the H_∞ framework requires

$$\max_{G \in \Pi} \| SW \|_\infty = \max_{G \in \Pi} \sup_\omega |S(j\omega)W(j\omega)| < 1, \tag{5.17}$$

where S is the sensitivity function, and W is the performance weight. In general, W^{-1} provides a bound on the maximum peak of the sensitivity function S, and imposes a minimum bandwidth constraint on the closed-loop.

In this section, the robust performance of the FSA system will be investigated. The following is a result for the robust performance of the FSA system.

Theorem 5.4   *Assume that the family of processes $\prod$ is described by (5.14). Furthermore, both $\frac{a(s)}{b(s)}$ and $\frac{\hat{a}(s)}{\hat{b}(s)}$ are strictly proper and they have the same number of unstable poles. Then the FSA system will meet the performance specification of (5.17) if*

$$|G_{yr}^*(j\omega)| \, \tilde{l}_m(\omega) + |(1 - G_{yr}^*(j\omega))W(j\omega)| < 1 \ , \quad \forall \omega. \tag{5.18}$$

PROOF.   The sensitivity function S is defined as

$$S(s) = 1 - G_{yr}(s).$$

Hence, it can be shown from (5.1) that

$$S(s) = \frac{1 + \frac{h(s)\hat{a}(s)}{p(s)q(s)} e^{-s\hat{L}}}{1 + \frac{h(s)\hat{a}(s)}{p(s)q(s)} e^{-s\hat{L}}(1 - \frac{a\hat{b}}{\hat{a}b} e^{-s(L-\hat{L})})},$$

which can be further simplified to

$$S(s) = \frac{1 - G_{yr}^*(s)}{1 - G_{yr}^*(s)l_m(s)}.$$

Replacing s by $j\omega$,

$$|S(j\omega)W| = \left| \frac{(1 - G_{yr}^*(j\omega))W(j\omega)}{1 - G_{yr}^*(j\omega)l_m(j\omega)} \right|$$

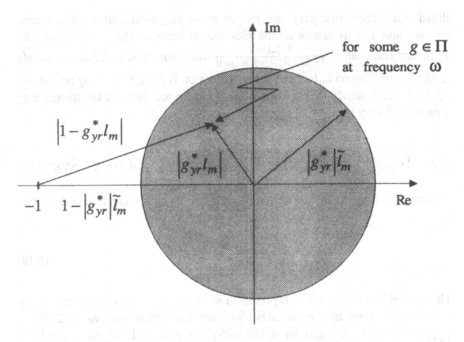

FIGURE 5.1. Geometrical construction for derivation of robust performance conditions.

Figure 5.1 shows a boundary plane on which $G_{yr}^* l_m$ is located for all $G \in \Pi$. For robust stability, it follows from (5.15) that $|G_{yr}^*(j\omega)|\tilde{l}_m(\omega) < 1$, $\forall\omega$. Thus, the boundary circular plane has a radius of less than 1 unit. From the construction, it is evident for $G \in \Pi$,

$$|1 - G_{yr}^*(j\omega)l_m(j\omega)| \geq 1 - |G_{yr}^*(j\omega)|\tilde{l}_m(j\omega), \forall\omega,$$

so that

$$\max_{G\in\Pi} |S(j\omega)W(j\omega)| \leq \frac{|(1 - G_{yr}^*(j\omega))W(j\omega)|}{1 - |G_{yr}^*(j\omega)|\tilde{l}_m(\omega)}.$$

It then follows that (5.17) is satisfied if

$$|G_{yr}^*(j\omega)|\tilde{l}_m(\omega) + |(1 - G_{yr}^*(j\omega))W(j\omega)| < 1 \quad, \quad \forall\omega.$$

and the proof is completed.

Remark 5.2   (5.18) can be rewritten as

$$\frac{|G_{yr}^*(j\omega)|}{1 - |(1 - G_{yr}^*(j\omega))W(j\omega)|} < \tilde{l}_m^{-1}(\omega).$$

Given the uncertainty bound $\tilde{l}_m(\omega)$ and the performance weight $W(j\omega)$, the robust performance design is thus to have a proper specification of the desired closed-loop polynomial $p(s)$ and the observer polynomial $q(s)$ while taking into account the dynamics of the closed-loop zeros so that the magnitude-frequency response of $\frac{|G_{yr}^*(j\omega)|}{1-|(1-G_{yr}^*(j\omega))W(j\omega)|}$ lies below that of $\tilde{l}_m^{-1}(\omega)$ to satisfy (5.18). Depending on $\tilde{l}_m(\omega)$ and the specification $W(j\omega)$, there may not be any $G_{yr}^*(j\omega)$ which satisfy (5.18), then $W(j\omega)$ may be too tight for the uncertainty present and may have to be relaxed.

5.4   Robust Controller Design for First-Order Systems

Assume that a model for the actual process has been obtained and is described by

$$\hat{G}_p(s) = \frac{\hat{K}}{\hat{T}s \pm 1}e^{-s\hat{L}}. \tag{5.19}$$

This model is well-known to represent a wide range of process dynamics typically encountered in the industry. Suppose the actual process is $G_p(s) = \frac{K}{Ts\pm1}e^{-sL}$, with top signs for a left-half-plane pole and bottom signs for a right-half-plane pole. Further, the parameters in the model are uncertain, *i.e.*

$$K_l \leq K \leq K_u, \quad T_l \leq T \leq T_u, \quad L_l \leq L \leq L_u,$$

with

$$\hat{K} = \frac{K_l + K_u}{2}, \quad \hat{T} = \frac{T_l + T_u}{2}, \quad \hat{L} = \frac{L_l + L_u}{2}.$$

The Uncertainty Bound   Given the simultaneous uncertainties in gain K, time-constant T and time-delay L, an exact analytical expression for the bound $\tilde{l}_m(\omega)$ is available (Laughlin *et al.* 1987) for both stable and unstable processes and it is given in (5.20)-(5.20).

$$\tilde{l}_m(\omega) = \begin{cases} \left|\left(\frac{|\hat{K}|+\Delta K}{|\hat{K}|}\right)\left(\frac{\pm j\hat{T}\omega+1}{j(\pm\hat{T}\mp\Delta T)\omega+1}\right)e^{(\pm j\Delta L\omega)} - 1\right|, & \forall \omega < \omega^*; \\ \left|\left(\frac{|\hat{K}|+\Delta K}{|\hat{K}|}\right)\left(\frac{\pm j\hat{T}\omega+1}{j(\pm\hat{T}\mp\Delta T)\omega+1}\right)\right| + 1, & \forall \omega \geq \omega^*, \end{cases} \tag{5.20}$$

where ω^* is defined implicitly by

$$\pm\Delta L\omega^* + \arctan\left[\frac{\pm\Delta T\omega^*}{1\pm\hat{T}(\pm\hat{T}\mp\Delta T)\omega^{*2}}\right] = \pm\pi, \quad \frac{\pi}{2} \leq \Delta L\omega^* \leq \pi, \tag{5.21}$$

and

$$\Delta K = |K_u - \hat{K}| < |\hat{K}|, \quad \Delta T = |T_u - \hat{T}| < |T_o|, \quad \Delta L = |L_u - \hat{L}| < |\hat{L}|.$$

When only the time-delay $\hat{L}$ is uncertain, *i.e.* $\Delta T = \Delta K = 0$, then $\omega^* = \frac{\pi}{\Delta L}$ and the above general expression simplifies to

$$\tilde{l}_m(\omega) = \left\{ \begin{array}{ll} \left| e^{(\pm j \Delta L \omega)} - 1 \right|, & \forall \omega < \frac{\pi}{\Delta L}; \\ 2, & \forall \omega \geq \frac{\pi}{\Delta L}. \end{array} \right.$$

When only the gain $\hat{K}$ is uncertain, *i.e.* $\Delta T = \Delta L = 0$, the expression simplifies to

$$\tilde{l}_m(\omega) = \frac{\Delta K}{|\hat{K}|}, \quad \forall \omega.$$

Given the uncertainty bound $\tilde{l}_m$, the closed-loop polynomial $p(s)$ and observer polynomial $q(s)$ may then be chosen to satisfy the robustness conditions.

Example 5.2   Assume that a nominal model is described by

$$\hat{G}_p(s) = \frac{0.8}{s+1} e^{-5s},$$

and all 3 parameters of the model are uncertain,

$$0.7 \leq K \leq 0.9, \quad 0.8 \leq T \leq 1.2, \quad 4 \leq L \leq 6.$$

It is desired to design the FSA system for robust performance with the performance weight as $W^{-1} = 2$. Choosing the damping factor as $\zeta = 1$ and the natural frequency as $\omega_o = 0.32$, (5.18) is satisfied since the magnitude-frequency response of $\frac{|G_{yr}^*(j\omega)|}{1 - |(1 - G_{yr}^*(j\omega))W(j\omega)|}$ lies below that of $\tilde{l}_m^{-1}(\omega)$ as shown in Figure 5.2. The FSA system is thus commissioned.

The performance of the FSA system to a unit step change of the reference signal at $t = 0$ and a 10% load disturbance at $t = 50$ is shown in Figure 5.3 for the extreme case perturbation in the process parameters of $K = 0.7$, $T = 0.8$ and $L = 6$. Clearly, in spite of the perturbed conditions, the proposed system demonstrates a performance robustness with tight control of the process variable towards asymptotic set-point tracking and load disturbance rejection, as opposed to the performance of a FSA system designed without robustness considerations.

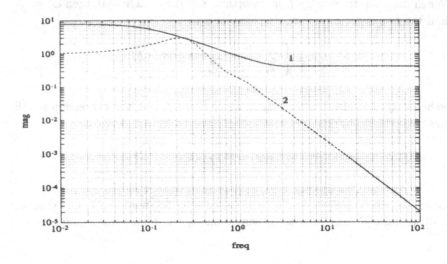

FIGURE 5.2. The magnitude-frequency response of (1) $\tilde{l}_m^{-1}(\omega)$ and (2) $\dfrac{|G_{yr}^{\circ}(j\omega)|}{1-|(1-G_{yr}^{\circ}(j\omega))W(j\omega)|}$ – Example 5.2.

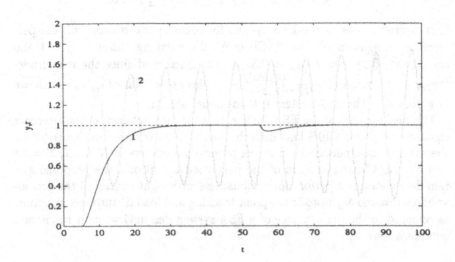

FIGURE 5.3. Performance of (1) the proposed robust FSA system, (2) a non-robust FSA system – Example 5.2.

5.5   Problems

1. Consider a process described by the following transfer function :

$$G(s) = \frac{Y(s)}{U(s)} = \frac{1}{3s+1}e^{-s}.$$

Design a FSA control system to maintain robust stability for an allowable 10% deviation in the model parameters.

2. Consider a process described by the following transfer function :

$$G(s) = \frac{Y(s)}{U(s)} = \frac{0.0001s^2 + 0.0002s + 1}{(0.0001s + 1)(s + 1)}e^{-s}.$$

Using the following model for the actual process :

$$\hat{G}(s) = \frac{Y(s)}{U(s)} = \frac{1}{s+1}e^{-s},$$

write a simulation program to implement the nominal FSA control system. Comment on the closed-loop stability of the system.

3. What is the difference between practical and robust stability?

4. Derive the practical stability results in Section 5.1 from the robust stability results in Section 5.2 ?

5. When will practical instability occurs with respect to properness of process and its model?

6. What can you say about stability robustness if the process has a different number of unstable poles from its model?

References

Alevisakis, G., and Segorg, D. E.(1973). An extention of the Smith Predictor method to multi-variable linear systems containing time delay. *Int. J. Control*, **3**, 541.

Artstein, Z. (1982). Linear systems with delayed controls: a reduction. *IEEE. Trans. Autom. Control*, **27**, 869.

Astrom, K. J. (1982). Ziegler-Nichols auto-tuners. *Internal Report TFRT-3167*. Dept of Automatic Control, Lund Institute of Technology, Lund, Sweden.

Astrom, K. J. (1991). Assessment of achievable performance of simple feedback loops. *Int Journal of Adaptive Control and Signal Processing*, **5**, 3-190.

Astrom, K. J. (1993). Autonomous controllers. *Control Eng. Practice*, **1**(2), 227-232.

Astrom, K. J. and Hagglund, T. (1984a). Automatic tuning of simple regulators. *Proceedings of the 9th IFAC World Congress*, Budapest, 1867–1872.

Astrom, K. J, and Hagglund, T. (1984b). Automatic tuning of simple regulators with specifications on phase and amplitude margins. *Automatica*, **20**(5), 645–651.

Astrom, K. J, and Hagglund, T. (1988a). *Automatic tuning of PID controllers*. Instrument Society of America, NC, USA.

Astrom, K. J, and Hagglund, T. (1988b). A new auto-tuning design. *Preprints IFAC Int. Symposium on Adaptive Control of Chemical Processes, ADCHEM '88*, Lyngby, Denmark, 141–146.

Astrom, K. J, and Wittenmark, B. (1989). *Adaptive Control*. Addison-Wesley:Reading, MA.

Astrom, K. J., Hagglund, T., Hang, C. C. and Ho, W. K. (1992a) Automatic tuning and adaptation for PID controllers - A survey. *Proceedings of the IFAC ACASP '92*, Grenoble, 121–126.

Astrom, K. J., Hang, C. C., Persson, P. and Ho, W. K. (1992b) Towards intelligent PID control. *Automatica*, **28**(1), 1–9.

Astrom, K. J. and Hagglund, T. (1995).*PID Controllers: Theory, Design, and Tuning*, 2nd Edition. Instrument Society of America.

Astrom, K. J., Lee, T. H, Tan, K. K. and Johansson, K. H. (1995). Recent advances in relay feedback methods – a survey. *Proceedings of the IEEE ICMSC'95*, Vancouver.

Astrom, K. J. and Wittenmark, B. (1989). *Adaptive control*. Addison Wesley, Book Company, Reading, Massachusetts, USA.

Cartwright, M. (1990). *Fourier Methods for Mathematicians, Scientists and Engineers*. Ellis Horwood.

Chiang, R.C., and C. C. Yu (1993). Monitoring Procedure for Intelligent Control: On- Line Identification of Maximum Closed-loop Log Modulus. *Ind. Eng. Chem. Res.*, **32**, 90.

DataTranslation (1995). *Building an Operator Interface in DT VEE, User Manual*; Data Translation, Inc.

Doyle, J. C. and Stein, G. (1981). Multivariable feedback design : concepts for a classical/modern synthesis. *IEEE Trans. Aut. Control*, **AC-26**, 4–16.

Friman, M., and K. Waller (1994). Auto-tuning of Multi-loop Control System. *Ind. Eng. Chem. Res.*, **33**, 1708.

Fisher Controls International, (1992). *DPR900 Instruction Manual*. Austin, Texas, USA.

Furukawa, F, and Shimemura, E. (1983). Predictive control for systems with delay. *Int. J. Control*, **37**, 399–412.

Gawthrop, P. J. (1977). Some interpretation of self-tuning controllers. *Proc. IEE*,**124**, 889.

Hagglund, T, and Astrom, K. J. (1991). Industrial adaptive controllers based on frequency response techniques. *Automatica*, **27**, 599–609.

Halevi, Y, (1991). Optimal reduced order models with delay. *Proceedings of the 30th Conference on Decision and Control*, Brighton, England, 602–607.

Hang, C. C, and Chin, D. (1991). Reduced order process modelling in self-tuning control. *Automatica*, **27**(3), 529–534.

Hang, C. C., Astrom, K. J. and Ho, W. K. (1993a). Relay auto-tuning in the presence of static load disturbance. *Automatica*, **29**(2), 563–564.

Hang, C. C, Lee, T. H.; Ho, W. K.(1993b). *Adaptive Control*; Instrument Society of America.

Hang, C. C, Wang, Q. G.; Zhou. J. H.(1994). Automatic Process Modelling from Relay Feedback. *Proc. IFAC Symp. on System Ident.*, **2**, 285-290.

Hang, C. C., Wang, Q. G.; Cao, L. S.(1995). Self-tuning Smith Predictors for Processes with Long Dead Time. *Int. J. of Adaptive Control and Signal Processing*, **9**, 255-270.

Haykin, S., (1989). *An Introduction to Analog & Digital Communications;* John Wiley & Sons.

Ho, W. K., Hang, C. C. and Cao, L. S. (1993). Tuning of PID controllers based on gain and phase margin specifications. *Preprints of the 12th IFAC World Congress,* **5**, 267-270.

Hocken, R. D., Salehi, S. V. and Marshall, J. E. (1983). Time-delay mismatch and the performance of predictor control schemes. *Int. J. Control,* **38**(2), 433-447

Holmberg, U, (1991). Relay feedback of simple systems. *Doctoral Dissertation,* Department of Automatic Control, Lund Institute of Technology.

Honeywell (1995). *Robust Multi-variable Predictive Control Technology, Robust PID.* Honeywell Industrial Automation and Control.

Huang, H. P, Chen C. L., Lai C. W. and Wang G. B. (1996). Auto-tuning for Model-Based PID Controllers. *AIChE Journal,* **42** (9), 2687-2691.

Ichikawa, K. (1985). Frequency-domain pole assignment and exact model-matching for delay systems. *Int. J. Control,* **41**, 1015-1024.

Kamen, E. W. (1982). Linear systems with commensurate time delays: stability and stabilization independent of delay. *IEEE. Trans. Autom. Control,* **27**, 367-375.

KentRidge (1992).*Dual Process Simulator KI 100, User Guide;* KentRidge Instruments Pte Ltd.

Kraus, T. W. and Myron, T. J. (1984). Self-tuning PID controller uses pattern recognition approach. *Control Engineering,* 108-111.

Krishnaswamy, P. R., B. E. M. Chan and G. P. Rangaiah (1987). Closed-loop Tuning of Process Control Systems.*Chem. Eng. Sci.,* **42**, 2173.

Kuhfittig, P. K. F. (1978). *Introduction To The Laplace Transform.* Plenum Press.

Jutan, A. and Rodriguez II, E. S. (1984). Extension of a New Method for on-line Controller Tuning, *Can. J. Chem. Engng,* **62**, 802-807.

Lee, T. H., Wang, Q. G., Tan, K. K. (1995). A knowledge-based Approach to Dead-time Estimation for Process Control. *Int. J. Control,* **61** (5), 1045-1072.

Lee, T. H., Wang, Q. G. and Tan, K. K. (1995a). Knowledge-based process identification from relay feedback. *Journal of Process Control*, 5(6), 387–397.

Lee, T. H., Wang, Q. G. and Tan, K. K. (1995b). Automatic tuning of the Smith predictor controller. *J. Syst. Engrg.*, 5(2), 102–114.

Li, W., Eskinat, E.; Luyben, W. L.(1991). An Improved Autotune Identification Method. *Ind. Eng. Chem. Res.*, 30, 1530-1541.

Lin, Y. J., and Yu C. C. (1993). Automatic Tuning and Gain Scheduling for pH Control. *Chem. Eng. Sci.*, 48, 3159.

Ljung, L, (1987). *System identification—theory for the user*. Prentice-Hall, Englewoods-Cliff.

Lundh, M, (1991). Robust adaptive control. *PhD thesis*, Lund Institute of Technology, Lund, Sweden.

Luyben, W. L, (1987). Derivation of Transfer Functions for Highly Nonlinear Distillation Columns. *Ind. Eng. Chem. Res.*, 26, 2490-2495.

Luyben, W. L, (1990) Process Modeling, Simulation and Control for Chemical Engineers; 2nd ed.; *McGraw-Hill: New York*.

Manitius, A. Z, and Olbrot, A. W. (1979). Finite spectrum assignment problem for systems with delays. *IEEE Trans. Aut. Control*, **AC-24**, 541–553.

Marshall, J. E. (1979). *Control of delay systems*. Stevenage, Peter Peregrimus Ltd, London.

Marshall, S. A. (1980). The design of reduced order systems. *Int. J. Control*, 30, 677–690.

Morari, M, and Zafiriou, E. (1989). *Robust process control*. Englewood Cliffs, NJ, Prentice Hall.

Murrill, P. W. (1988). *Application concepts of process control*. Instrument Society of America Press, Research Triangle Park, NC.

Ogunnaike, B.A. and Ray, W.H. (1979).*J. Am. Inst. Chem. Engrs*, 25, 1043.

Olbrot, A.W. (1981). *IEEE. Trans. Autom. Control*,26, 513.

Palmor, Z. J, (1980). Stability properties of Smith dead-time compensator controllers. *Int. J. Control*, 53, 937–949.

Palmor, Z. J, (1982). Properties of optimal stochastic control systems with dead-time. *Automatica*, 18(1), 107–116.

Palmor, Z. J, and Blau, M. (1994). An auto-tuner for Smith dead-time compensators. *Int. J. Control*, 60(1), 117–135.

Palmor, Z. J. and Halevi, Y. (1983). On the design and properties of multi-variable dead time compensation. *Automatica,* **19**, 255–264.

Palmor, Z. J. and Powers, D. V. (1985). Improved dead-time compensator controllers. *AIChE Journal,* **31**(2), 215–221.

Palmor, Z. J. and Shinnar, R. (1981). Design of advanced process controllers. *AIChe Journal,* **27**(5), 793–805.

Pintelon, et al. (1994). Parametric Identification Of Transfer Functions In Frequency Domain—A Survey. *IEEE Trans. on Automatic Control,* **39**, 2245–2259.

Ramirez, R. W, (1985). The FFT Fundamentals and Concepts; *Englewood Cliffs, N.J.: Prentice-Hall.*

Seborg, D. E., Edgar, T. F. and Mellichamp, D. A. (1989). *Process dynamics and control.* John Wiley & Sons Inc.

Shen, S. H., Wu, J. S.; Yu, C. C. (1996). Use of Biased-Relay Feedback for System Identification. *AICHE Journal.* **42**, 1174-1180.

Slotine, J-J. E.　and Li, W. P. (1991). *Applied nonlinear control.* Prentice Hall, Englewoods Cliffs, NJ, USA.

Smith, O. J. M, (1957). Closer control of loops with dead-time. *Chemical Engineering Progress,* **53**, 217–219.

Smith, O. J. M. (1959). A controller to overcome dead-time. *ISA J.,* **6**, 28–33.

Tan, K. K., Lee, T. H. and Wang, Q. G. (1996). Robustness of finite spectrum-assigned delay systems, *Control Theory and Advanced Technology,* **10**(4), Part 4.

Tan, K. K., Wang, Q. G. and Lee, T. H. (1996). Enhanced automatic tuning procedure for PI/PID controllers for process control. *AIChE Journal,* **42**(9), 2555-2562.

Wang, Q. G. (1984), Modelling and control of papermaking processes. *M.S.thesis,* Zhejiang University.

Wang, Q. G., Sun Y. X. and Zhou, C. H. (1988). Finite spectrum assignment for multivariable delay systems in frequency domain. *Int. J. Control,* **47**(3), 729–734.

Wang, Q. G., Lee, T. H. and Hang, C. C. (1993). Frequency-domain finite spectrum assignment for delay systems with multiple poles. *Int. J. Control,* **58**, 735–738.

Wang, Q. G., Lee, T. H. and Hang, C. C. (1993). Practical stability of delay systems with finite spectrum assignment. *Control Theory and Advanced Technology,* **10**(4), Pt 1, 913–922.

Wang, Q. G., Lee, T. H. and Tan, K. K. (1995). Automatic tuning of finite spectrum assignment controllers for delay systems. *Automatica*, **31**(3), 477–482.

Wang, Q. G., Lee, T. H. and Tan, K. K. (1997). Finite spectrum assignment control of unstable time delay processes with relay tuning. *IECR*,**37**(4), 1351–1357.

Wang, Q. G., C. C. Hang and Q. Bi (1997). A Frequency Domain Controller Design Method. *Trans IChemE*, **75**, Part A, 64-72.

Wang, Z. Q. and Skogestad, S. (1993). Robust control of time-delay systems using the Smith predictor. *Int. J. Cont.*, **57**(6), 1405–1420.

Watanabe, K. and Ito, M. (1981). *IEEE Trans. Aut. Control*,**26**, 1261.

Watanabe, K., Nobuyama, E., Kitamori, T. and Ito, M. (1992). A new algorithm for finite spectrum assignment of single-input systems with time-delay. *IEEE Trans. Aut. Control*, **AC-37**(9), 1377–1383.

Wellstead, P. E. (1981). Non-Parametric Methods of System Identification. *Automatica*, **17**, 55-69.

Whitfield, A. H. (1986). Transfer Function Synthesis Using Frequency Response Data. *Int. J. Control*, **43**(5), 1413-1426.

Wolovich, W. A. (1974). *Linear Multivariable Systems*. Berlin: Springer-Verlag.

Yamanaka, K. and Shimemura, E. (1987). Effects of mismatched Smith controller on stability in systems with time delay. *Automatica*, **23**(6), 787–791.

Yuwana, M. and Seborg, D. E. (1982). A New Method for On-line Controller Tuning, *AIChE*, **28**, 434-440.

Ziegler, J. G, and Nichols, N. B. (1943). Optimum settings for automatic controllers. *Trans. ASME*, **65**, 433–444.

Lecture Notes in Control and Information Sciences

Edited by M. Thoma

1993–1998 Published Titles:

Vol. 186: Sreenath, N.
Systems Representation of Global Climate
Change Models. Foundation for a Systems
Science Approach.
288 pp. 1993 [3-540-19824-5]

Vol. 187: Morecki, A.; Bianchi, G.;
Jaworeck, K. (Eds)
RoManSy 9: Proceedings of the Ninth
CISM-IFToMM Symposium on Theory and
Practice of Robots and Manipulators.
476 pp. 1993 [3-540-19834-2]

Vol. 188: Naidu, D. Subbaram
Aeroassisted Orbital Transfer: Guidance
and Control Strategies
192 pp. 1993 [3-540-19819-9]

Vol. 189: Ilchmann, A.
Non-Identifier-Based High-Gain Adaptive
Control
220 pp. 1993 [3-540-19845-8]

Vol. 190: Chatila, R.; Hirzinger, G. (Eds)
Experimental Robotics II: The 2nd
International Symposium, Toulouse,
France, June 25-27 1991
580 pp. 1993 [3-540-19851-2]

Vol. 191: Blondel, V.
Simultaneous Stabilization of Linear
Systems
212 pp. 1993 [3-540-19862-8]

Vol. 192: Smith, R.S.; Dahleh, M. (Eds)
The Modeling of Uncertainty in Control
Systems
412 pp. 1993 [3-540-19870-9]

Vol. 193: Zinober, A.S.I. (Ed.)
Variable Structure and Lyapunov Control
428 pp. 1993 [3-540-19869-5]

Vol. 194: Cao, Xi-Ren
Realization Probabilities: The Dynamics of
Queuing Systems
336 pp. 1993 [3-540-19872-5]

Vol. 195: Liu, D.; Michel, A.N.
Dynamical Systems with Saturation
Nonlinearities: Analysis and Design
212 pp. 1994 [3-540-19888-1]

Vol. 196: Battilotti, S.
Noninteracting Control with Stability for
Nonlinear Systems
196 pp. 1994 [3-540-19891-1]

Vol. 197: Henry, J.; Yvon, J.P. (Eds)
System Modelling and Optimization
975 pp approx. 1994 [3-540-19893-8]

Vol. 198: Winter, H.; Nüßer, H.-G. (Eds)
Advanced Technologies for Air Traffic Flow
Management
225 pp approx. 1994 [3-540-19895-4]

Vol. 199: Cohen, G.; Quadrat, J.-P. (Eds)
11th International Conference on
Analysis and Optimization of Systems –
Discrete Event Systems: Sophia-Antipolis,
June 15–16–17, 1994
648 pp. 1994 [3-540-19896-2]

Vol. 200: Yoshikawa, T.; Miyazaki, F. (Eds)
Experimental Robotics III: The 3rd
International Symposium, Kyoto, Japan,
October 28-30, 1993
624 pp. 1994 [3-540-19905-5]

Vol. 201: Kogan, J.
Robust Stability and Convexity
192 pp. 1994 [3-540-19919-5]

Vol. 202: Francis, B.A.; Tannenbaum, A.R.
(Eds)
Feedback Control, Nonlinear Systems,
and Complexity
288 pp. 1995 [3-540-19943-8]

Vol. 203: Popkov, Y.S.
Macrosystems Theory and its Applications:
Equilibrium Models
344 pp. 1995 [3-540-19955-1]